SEASONS OF THE MOON

FOLK NAMES AND LORE OF THE FULL MOON

Michael Carabetta

CHRONICLE BOOKS
SAN FRANCISCO

Library of Congress Cataloging-in-Publication Data available.

ISBN 978-1-4521-7656-7

Manufactured in China.

MIX
Paper from responsible sources
FSC™ C008047
FSC www.fsc.org

Design: Katherine Yao.
Picture Editor: Julien Tomasello.

IMAGES: **1-2:** Shutterstock; **10-11:** © Michael Fuhr; **12:** © Rob Fullen; **14-15:** © Göran Strand; **17:** © David Sausse Photography; **18:** © George F Mobley/National Geographic; **20-21:** © Mike Shoys; **22:** © Dawn Richerson Photography; **24-25:** © DW Rose/Flickr/https://creativecommons.org/licenses/by/2.0/; **26-27:** © Arindam Dey/@arindamneoworld; **29:** © Sumit Nandy; **30:** © Tom Vezo/Minden Pictures; **32-33:** © Reiner Aichholz; **34-35:** © Hannele Lahti/National Geographic; **36:** © Anthony Devlin/Getty Images; **39:** © Norman McCloskey Photography; **40-41:** © Pramote Soongkitboon/Dreamstime.com; **43:** © Greg Redfern; **44-45:** © HDH Lucas; **46:** © Sergio Emilio Montúfar Codoñer/www.Pinceladasnocturnas.com; **49:** © Dan Cantlon; **50-51:** © Ronan Donovan/National Geographic; **52:** © Ashley/Getty Images; **55:** © 2014 Rory Sagner Photography & Fine Art/www.RorySagner.Pixels.com; **56-57:** © Juha-Pekka Kanto; **58:** © Juliana Lefterova; **61:** © Jeff Wizniak; **62-63:** © Ryo Iwasaki; **64-65:** © Aniket Thakur/Flickr/https://creativecommons.org/licenses/by/2.0/; **67:** © Ginger Parish; **68-69:** NASA/Lauren Hughes; **70:** giovigraf/EyeEm; **72-73:** © Javi Lorbada/www.Javilorbada.com; **74-75:** © Lennart Öhmark; **76:** © NASA/Bill Ingalls; **78-79:** Al Petteway & Amy White / National Geographic; **80-81:** © Rudy Calpo; **82-83:** © Stephen Fischer Photography; **84-85:** © NASA/Dominic Hart; **86-87:** © Dennis Doucet/DennisDoucet.MyPortfolio.com; **88-89:** © Fred Espenak / www.MrElipse.com; **92-93:** Shutterstock

10 9 8 7 6 5 4 3 2 1

Chronicle Books LLC
680 Second Street
San Francisco, California 94107
www.chroniclebooks.com

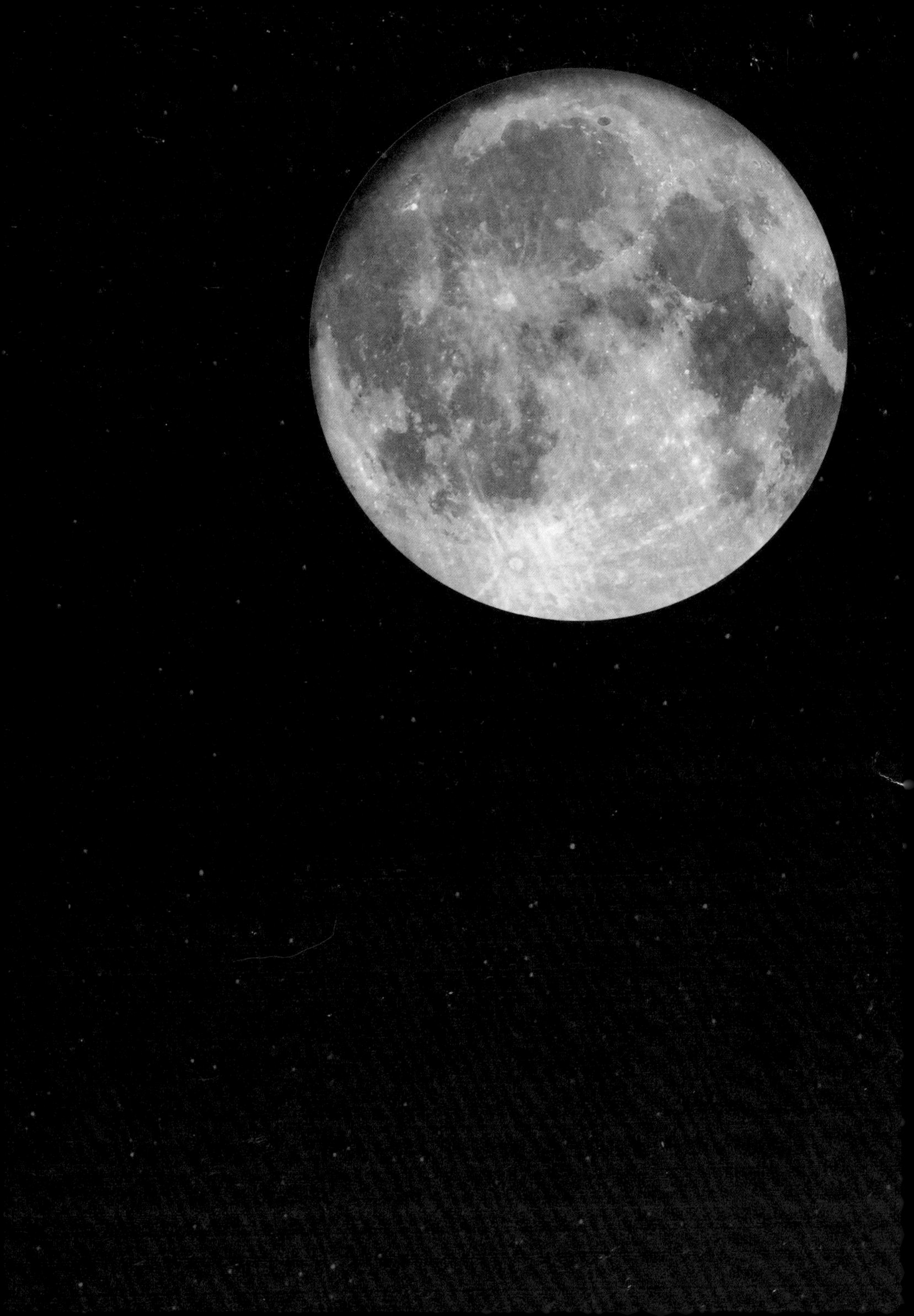

To everything there is a season,
and a time to every purpose under heaven:

Ecclesiastes 3:1–2

For
Julia, Elsinore & Bendix,
My forever sky marks.

CONTENTS

INTRODUCTION

THE MOON—ALWAYS THERE, always changing, tethered to our planet as we swing around the Sun.

It's a mystery. We want to understand it, and feel as though we know something about it. We feel a connection. We want to touch it. And in our way we have; we've landed on it, and brought some of it back home to Earth. It's both alien and familiar, close and yet so far away. It's a paradox, and has always been so.

Rising, it looms large, filling the horizon, yet on its journey across the sky (our journey, too), it diminishes until it sinks over the edge of the horizon or is masked by the Sun. It inspires our myths, fantasies, folklore, and, what's more, our art and our science.

Moonlight. It's eerie, unworldly, magic, liquid, silver . . . or is it? When eclipsed by Earth's shadow, the Moon appears blood red. It's called the Blood Moon.

I find the language we use to describe the Moon fascinating, especially the folk names it has inspired. Some of them I knew through their common usage, being the subject of a song or idiom: "shine on, harvest moon," or "once in a blue moon." But others were new to me, and piqued my interest.

"We will have a full moon on Thursday, a Buck Moon," I heard a radio announcer say one morning, as I made my breakfast. This name, I learned, was bestowed (like many others) by Native American tribes, who turned their attention to the full moons and gave them evocative titles that signified moments in time and season. The Buck Moon, for instance—occurring in July—marks the annual phase when male bucks typically begin to sprout new antlers in preparation for their tussles with each other in mating season. There's also a Strawberry Moon. And a Bone Moon.

In North America, where I live, the names that the regional native tribes here bestowed upon the full moons of the year are those most commonly heard. But of course, cultures around the world have been inspired by the Moon's mysterious beauty and presence throughout the seasons of their lives: Celtic, Chinese, Hindu, medieval English . . .

These names inspired me to look up at the Moon with a different perspective, and to explore some of the ways we have tried to touch its beauty and mystery in photographs from here on Earth. Each image in these pages was taken at the time of year of the full moon it represents.

On a clear night, the light of the moon, a reflection of the sun's rays, is inescapable. I hope this book captures some small sense of the wonder that we humans share for our celestial neighbor, in its light, through the dark, and across the changing times and seasons of our lives.

—Michael Carabetta

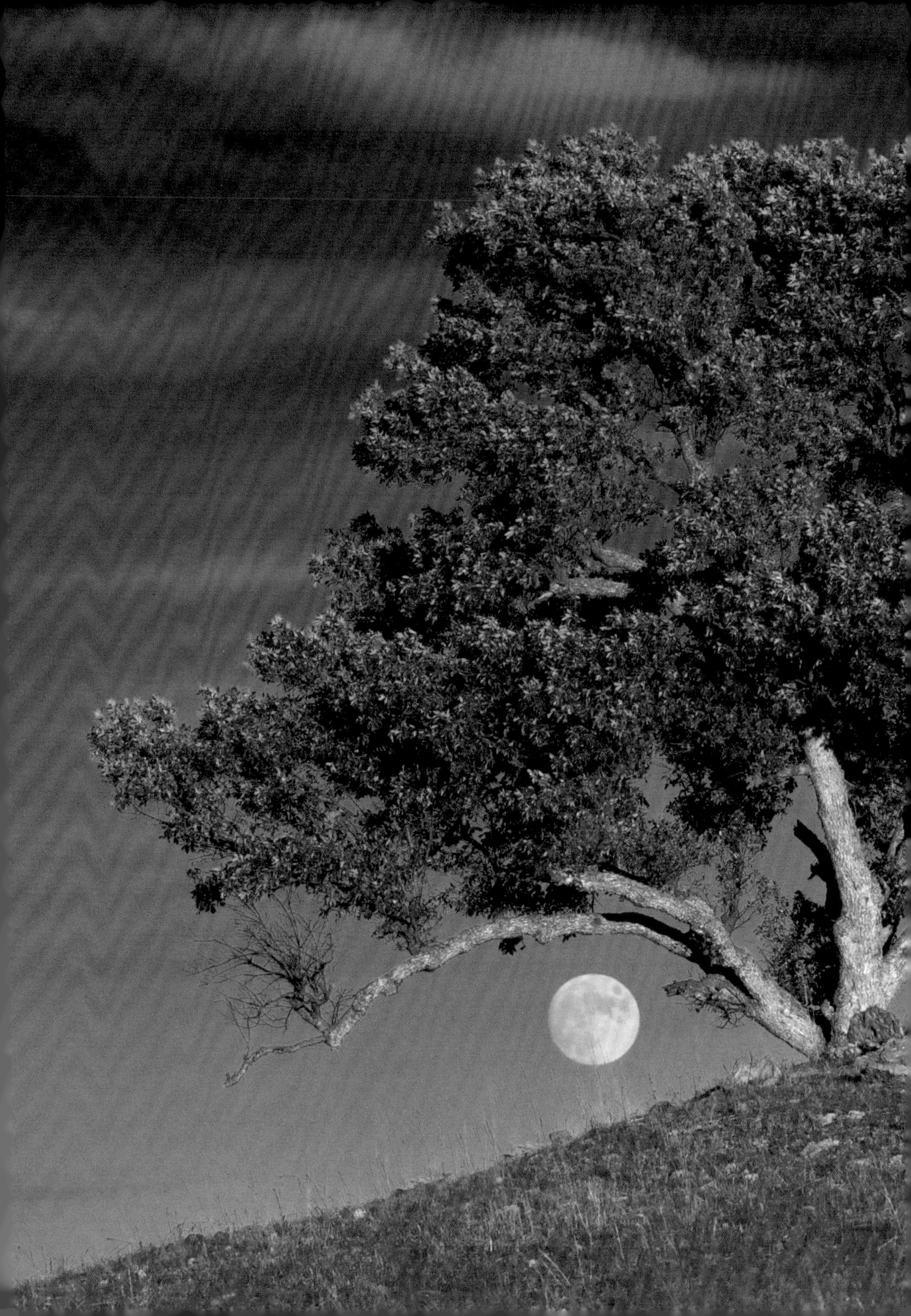

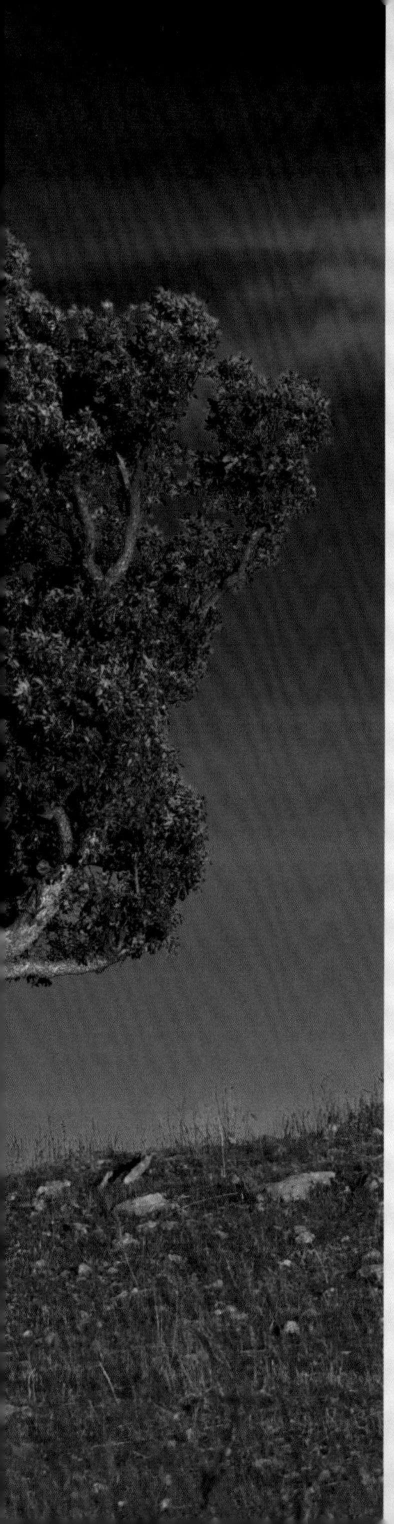

The moon, by her comparative proximity, and the constantly varying appearances produced by her several phases, has always occupied a considerable share of the attention of the inhabitants of the earth.

—Jules Verne

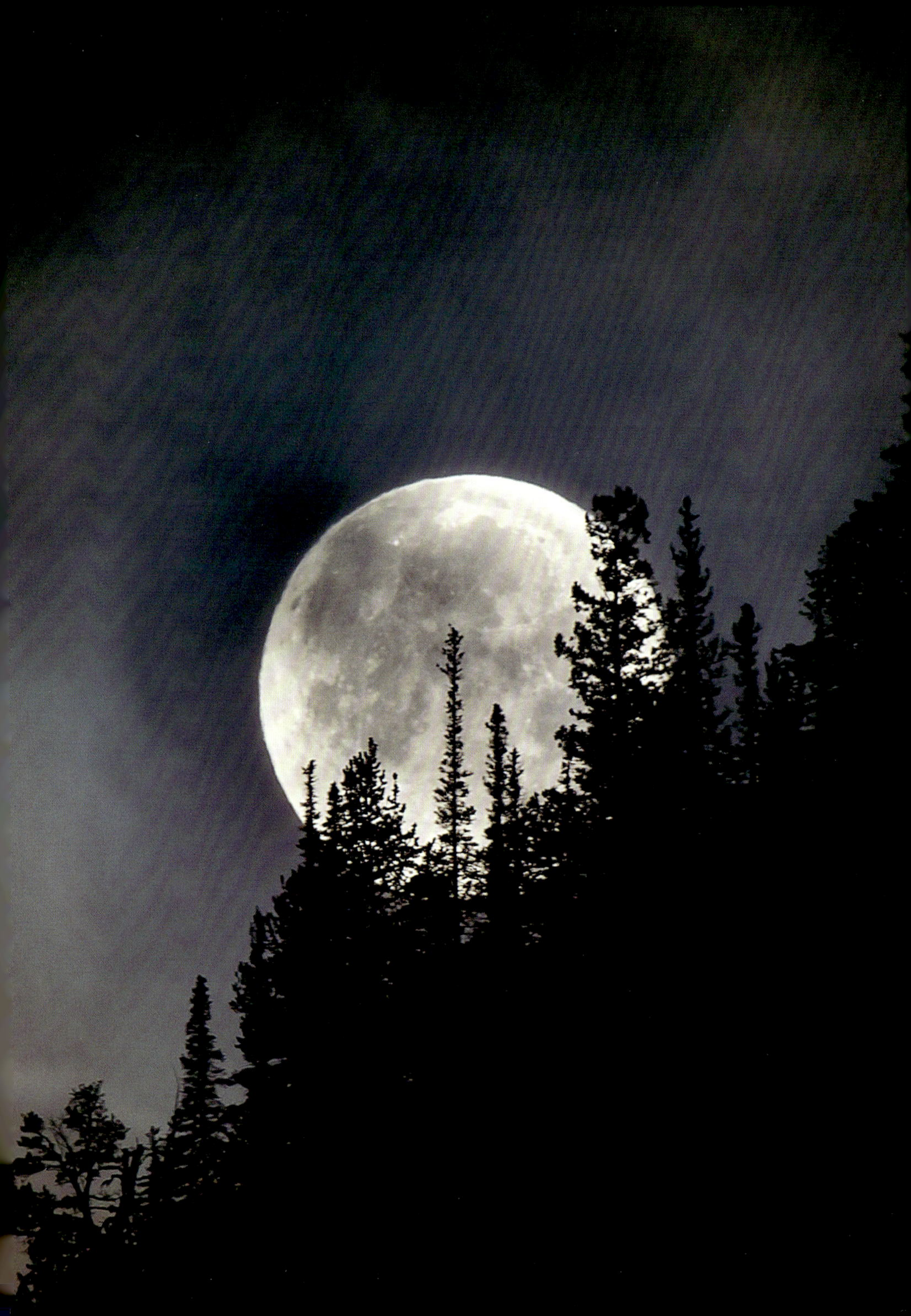

WOLF MOON

Medieval Europeans and a number of the Northern and Eastern Native American tribes alike were inspired to call January's full moon the Wolf Moon. Packs of wolves, hungry in winter, bayed loudly near human settlements, forever linking their howls to the first month of the year.

Like every great river and every great sea, the moon belongs to none and belongs to all. E.B. WHITE

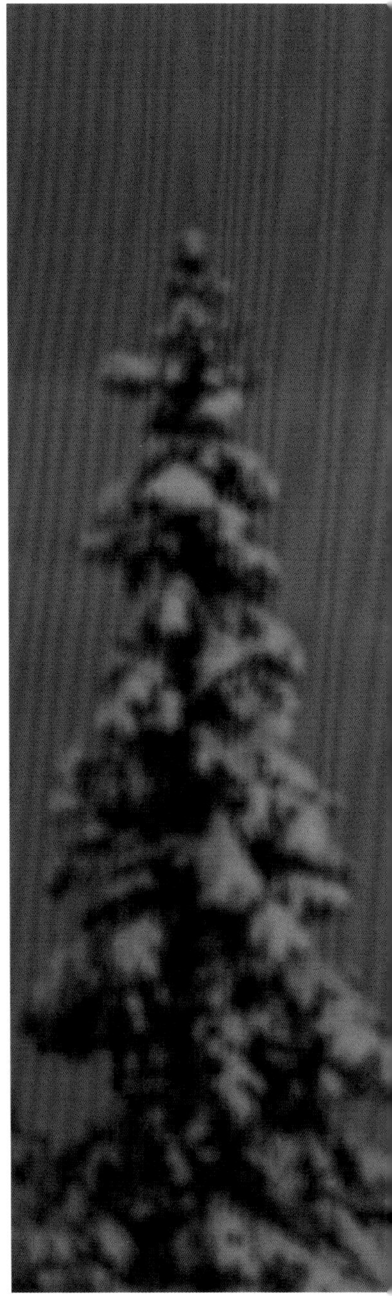

COLD MOON

The Lakota tribe called January's full moon the Moon of Hard Times. Life on the windswept and snow-blanketed prairies could be hard to endure, with fresh food in short supply during the severe winter months. The Cherokee people called it the Cold Moon to mark the arrival of winter's snow, ice, and frigid temperatures.

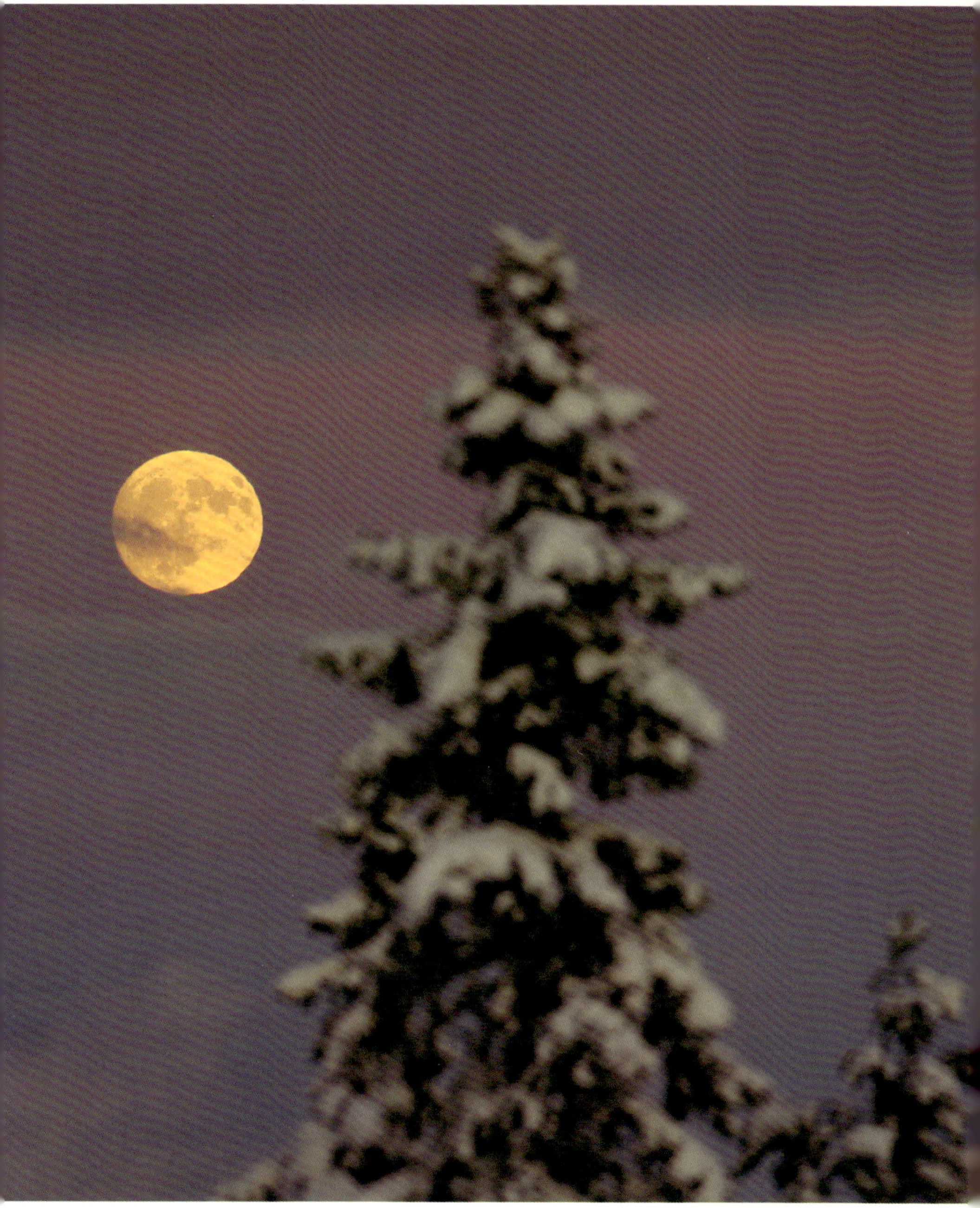

QUIET MOON

Among the Celtic people, the January full moon was known as the Quiet Moon, falling as it did on the quiet landscape—enveloped in snow.

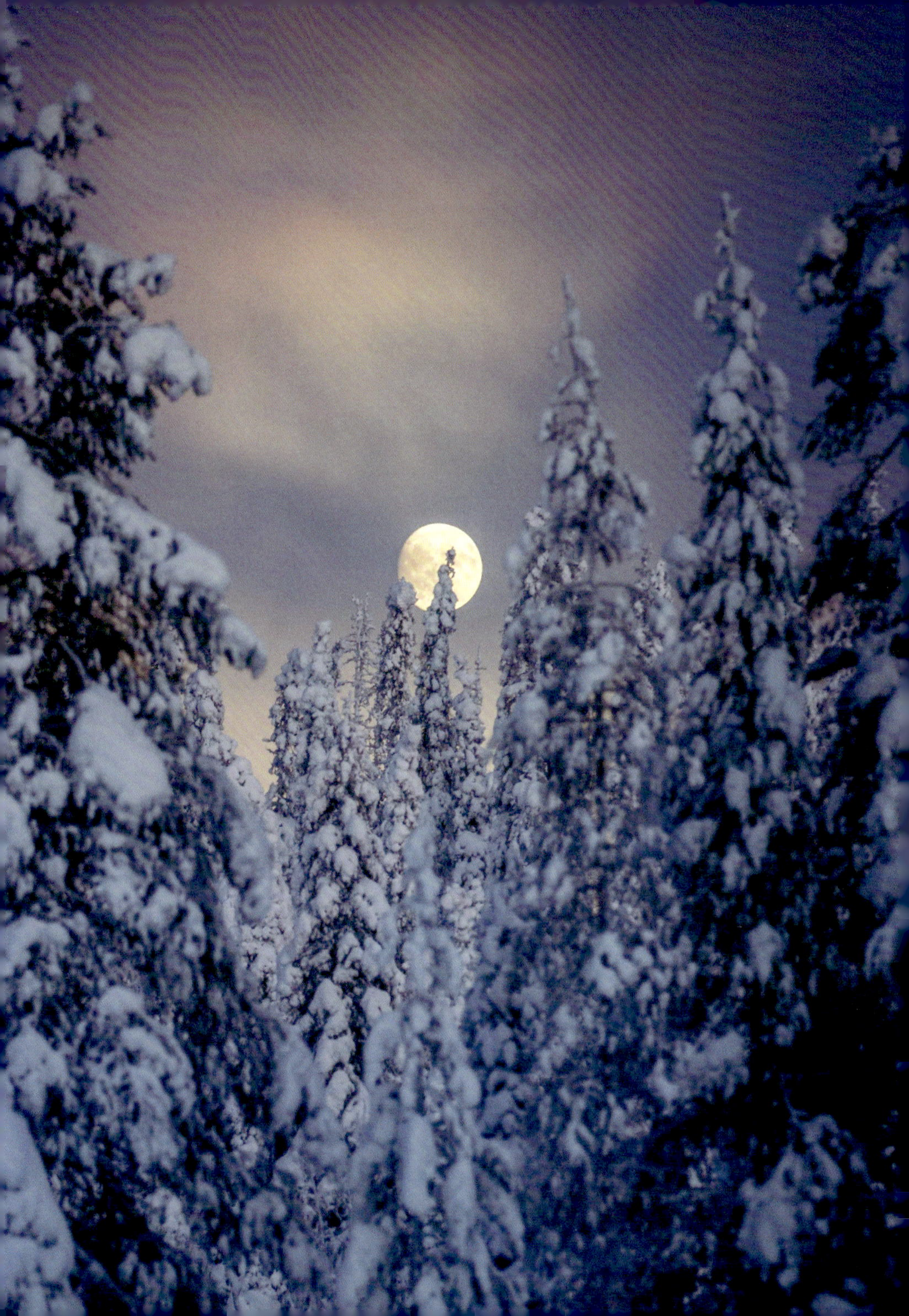

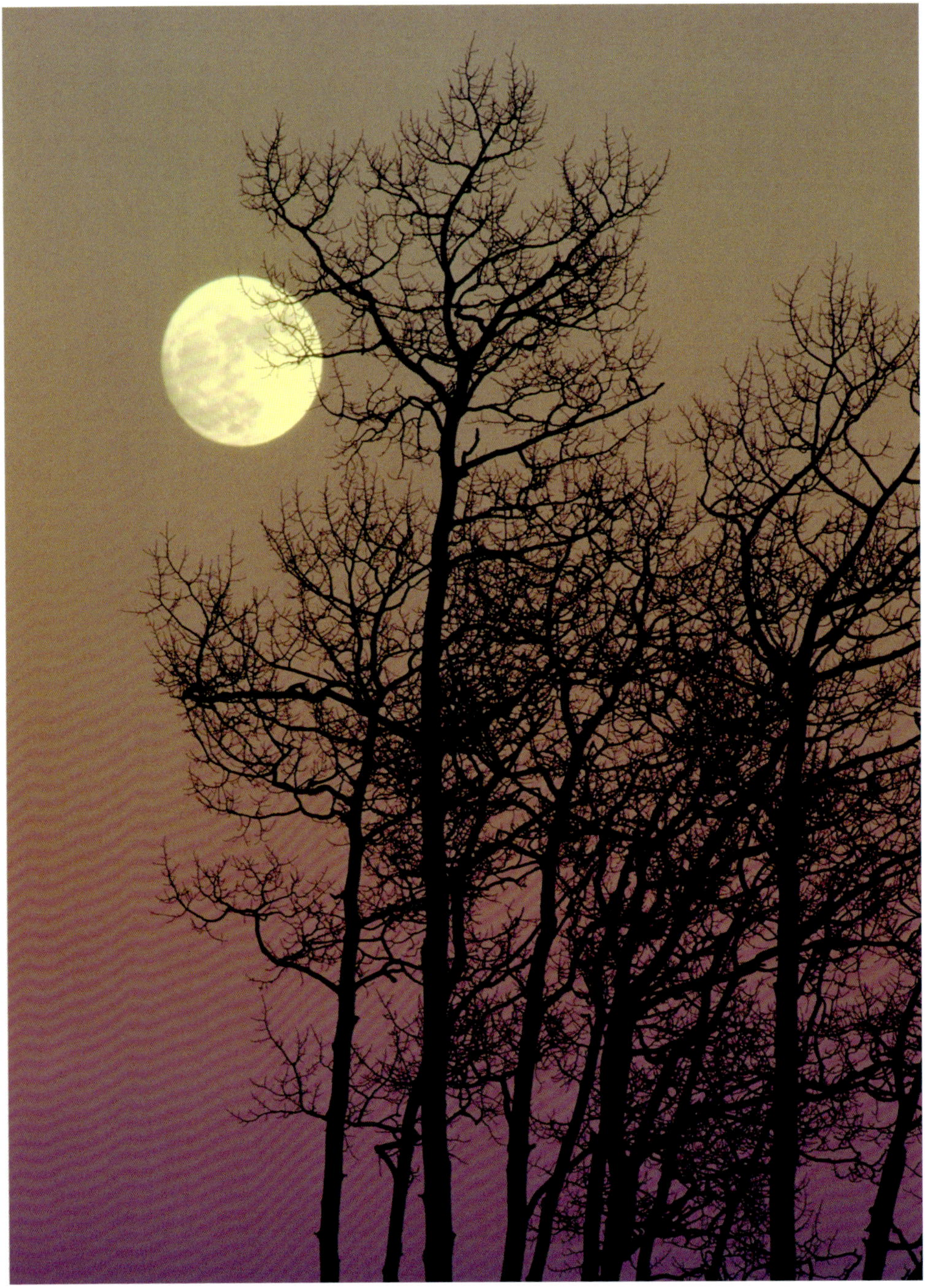

BUDDING MOON

In China, February's full moon became known as the Budding Moon. This was a time when sap was beginning to rise in the trees, a harbinger of when buds would start to form, ushering in spring.

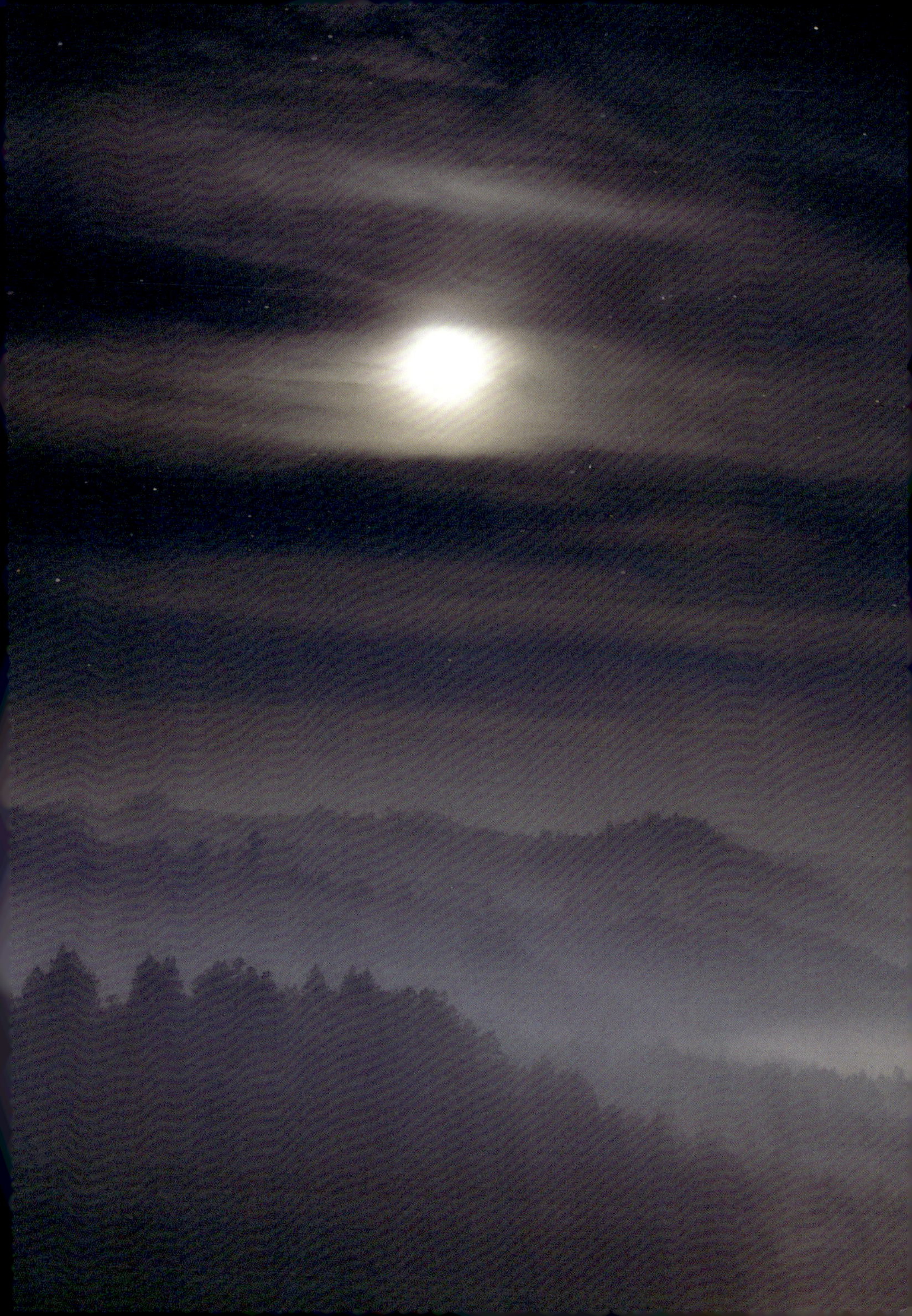

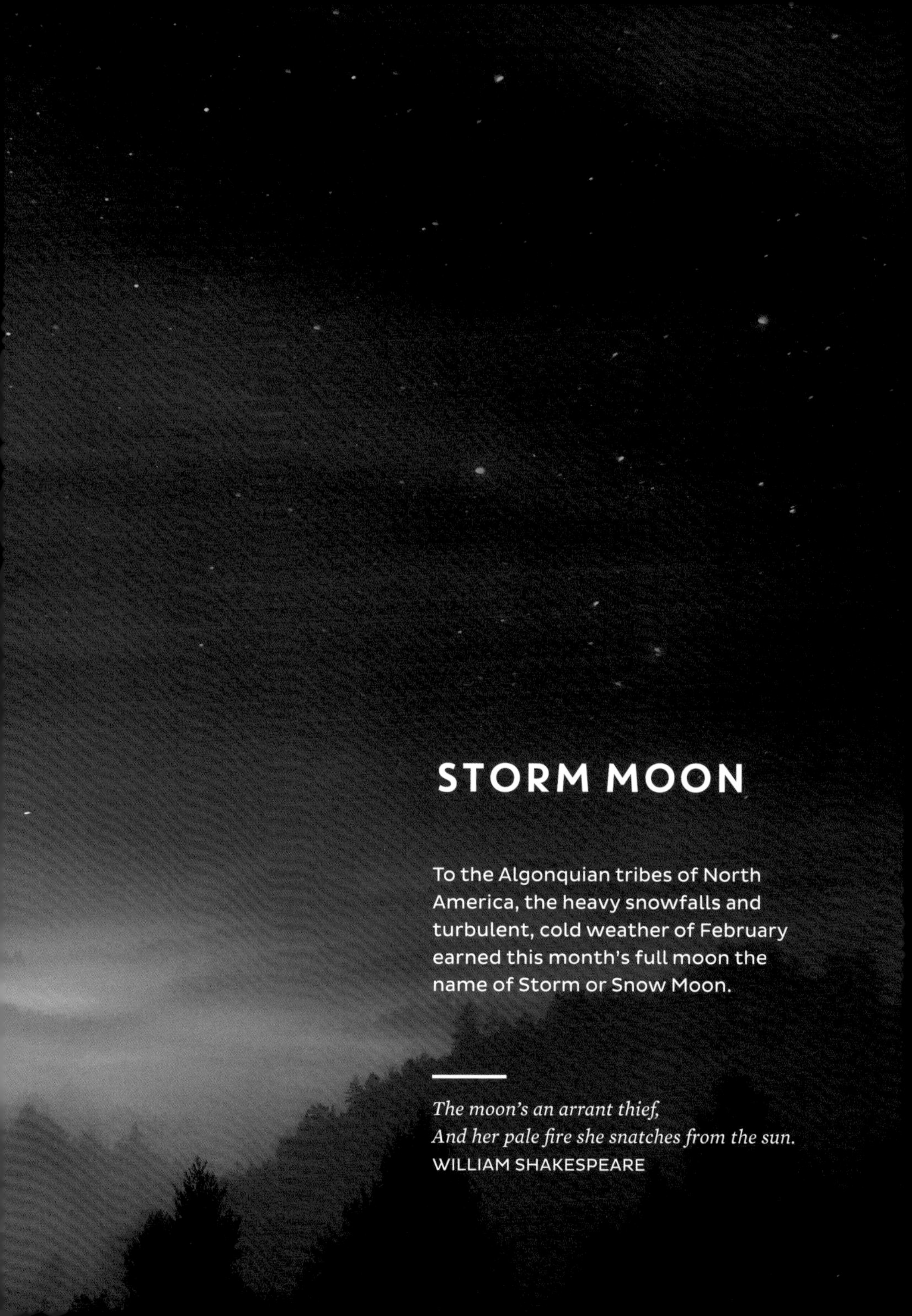

STORM MOON

To the Algonquian tribes of North America, the heavy snowfalls and turbulent, cold weather of February earned this month's full moon the name of Storm or Snow Moon.

The moon's an arrant thief,
And her pale fire she snatches from the sun.
WILLIAM SHAKESPEARE

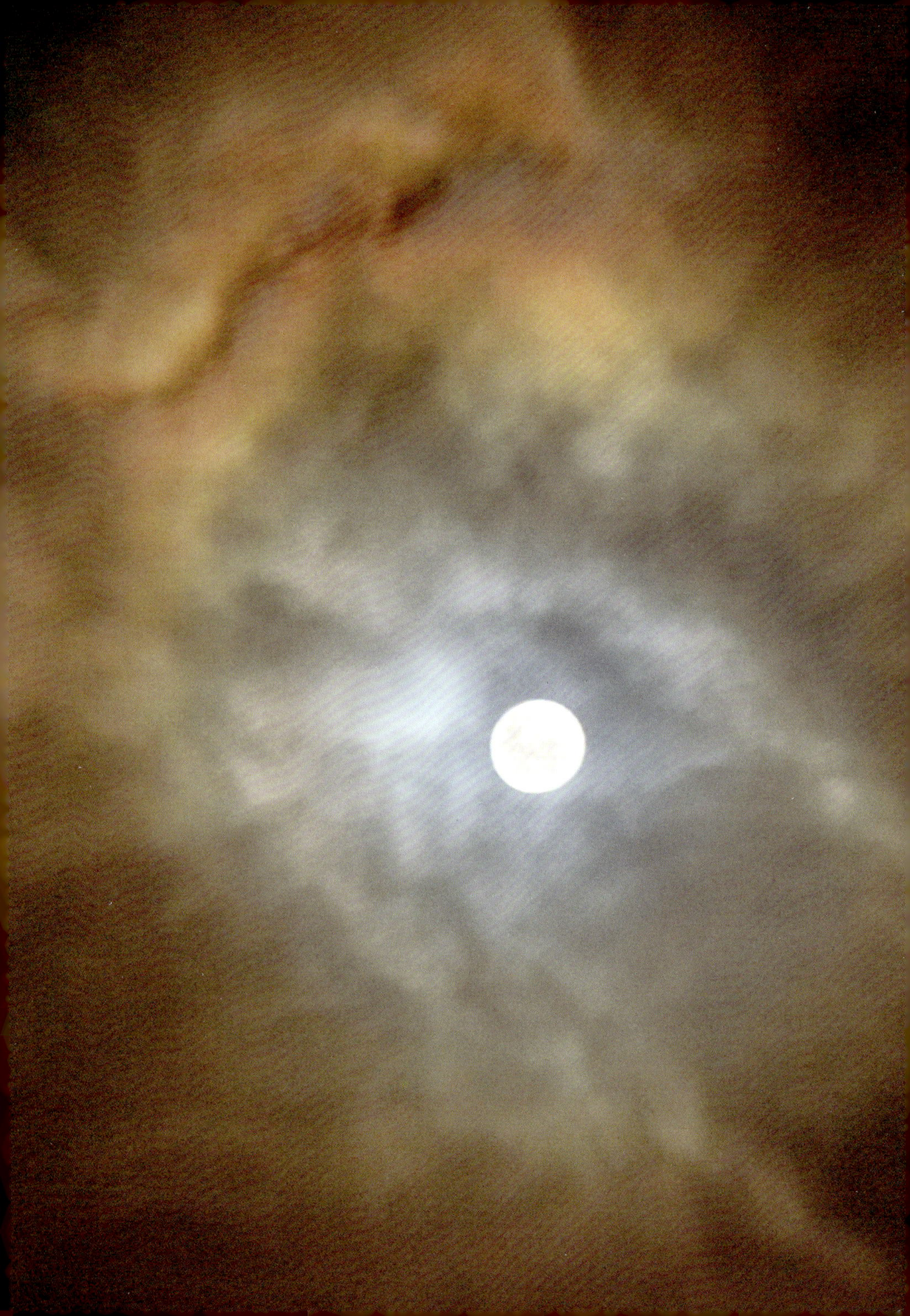

BONE MOON

The Cherokee people called this the Bone Moon, perhaps in part because of the bone-chilling temperatures at this time of year. But the name is more commonly attributed to food scarcity in this bleak winter month and the necessary resourcefulness of using every last scrap of food—including the bones—to make a meal.

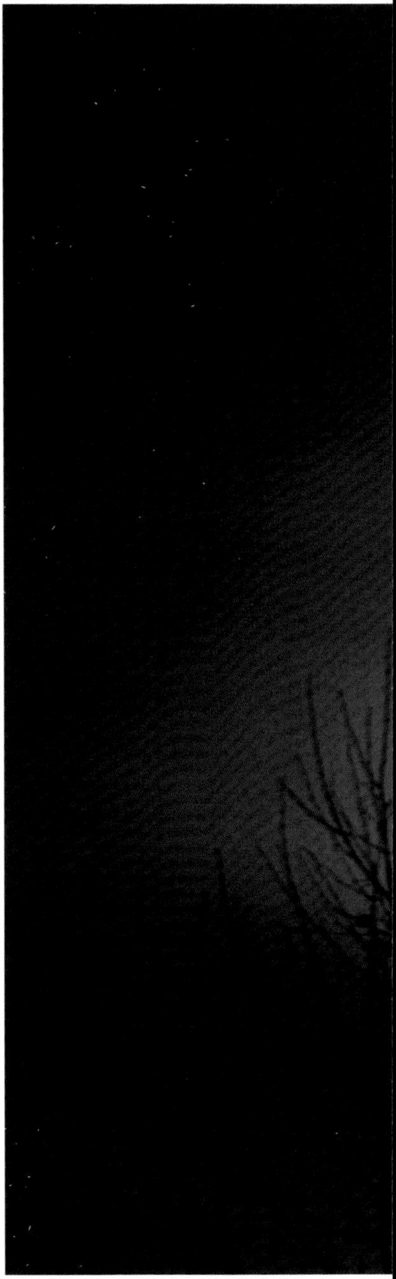

CHASTE MOON

In medieval England, March's full moon was known as the Chaste Moon, a reference to the earth as it thaws from the throes of winter, and was thought to lie fresh and pure, ready to begin the seasonal life cycle anew.

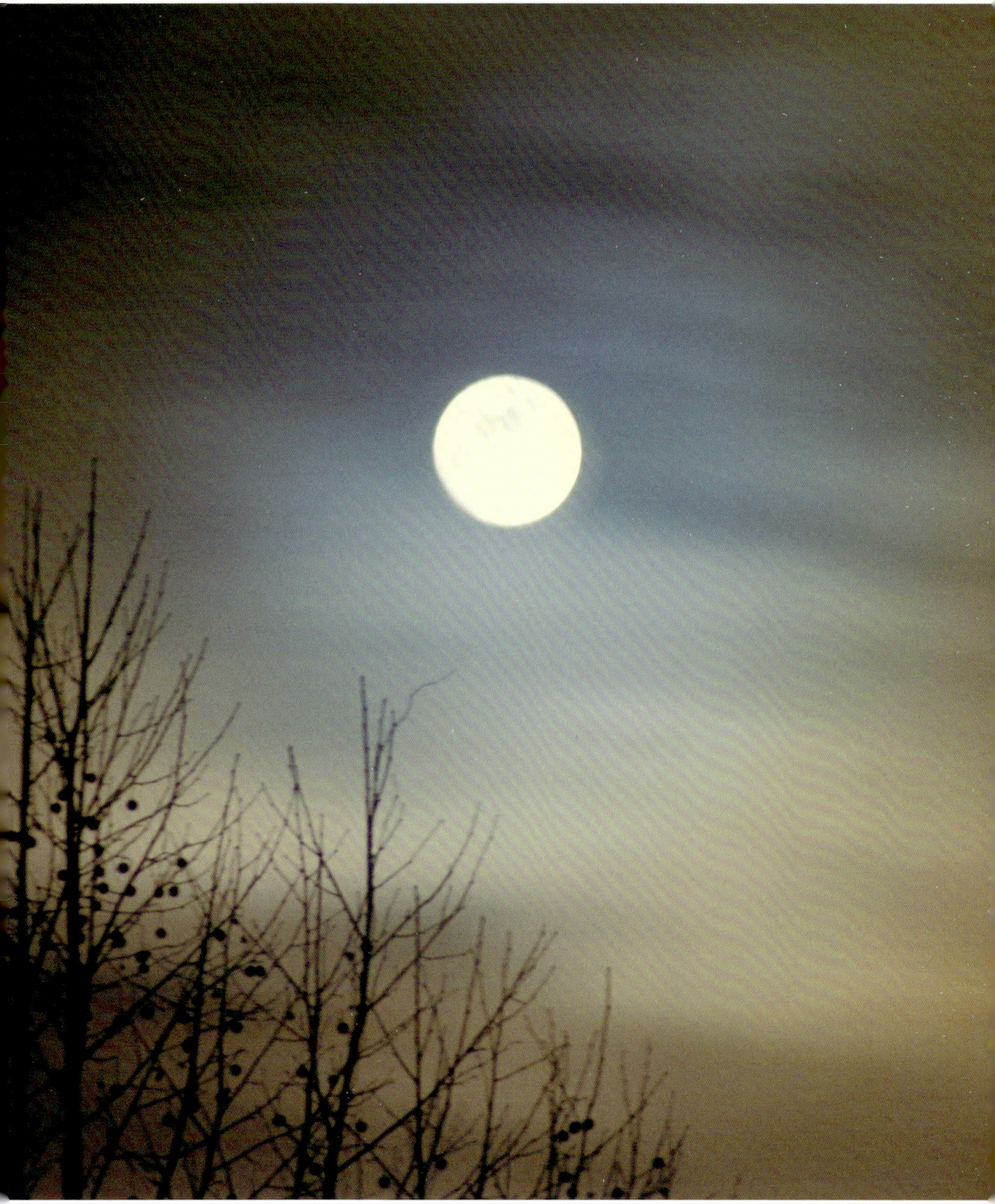

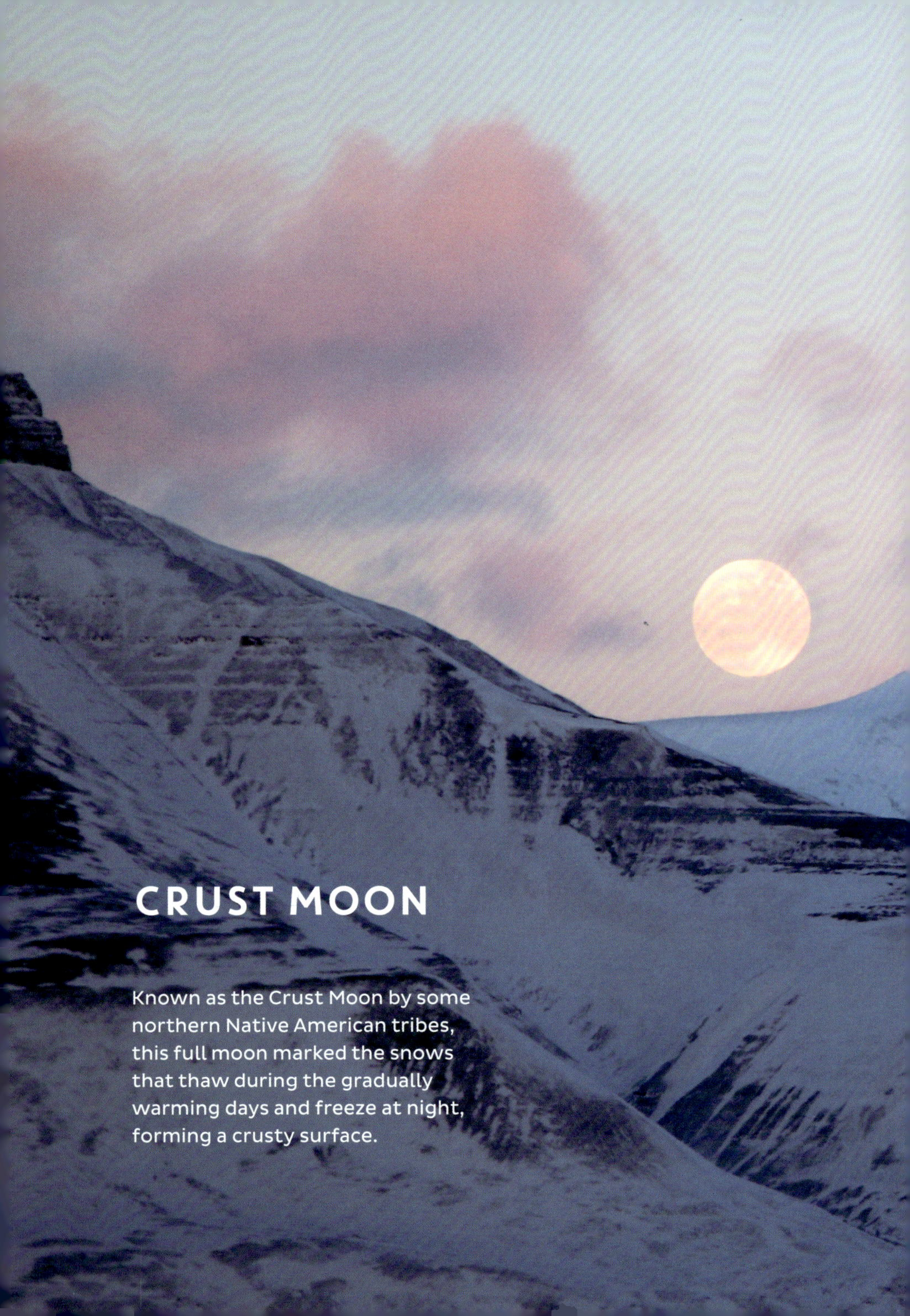

CRUST MOON

Known as the Crust Moon by some northern Native American tribes, this full moon marked the snows that thaw during the gradually warming days and freeze at night, forming a crusty surface.

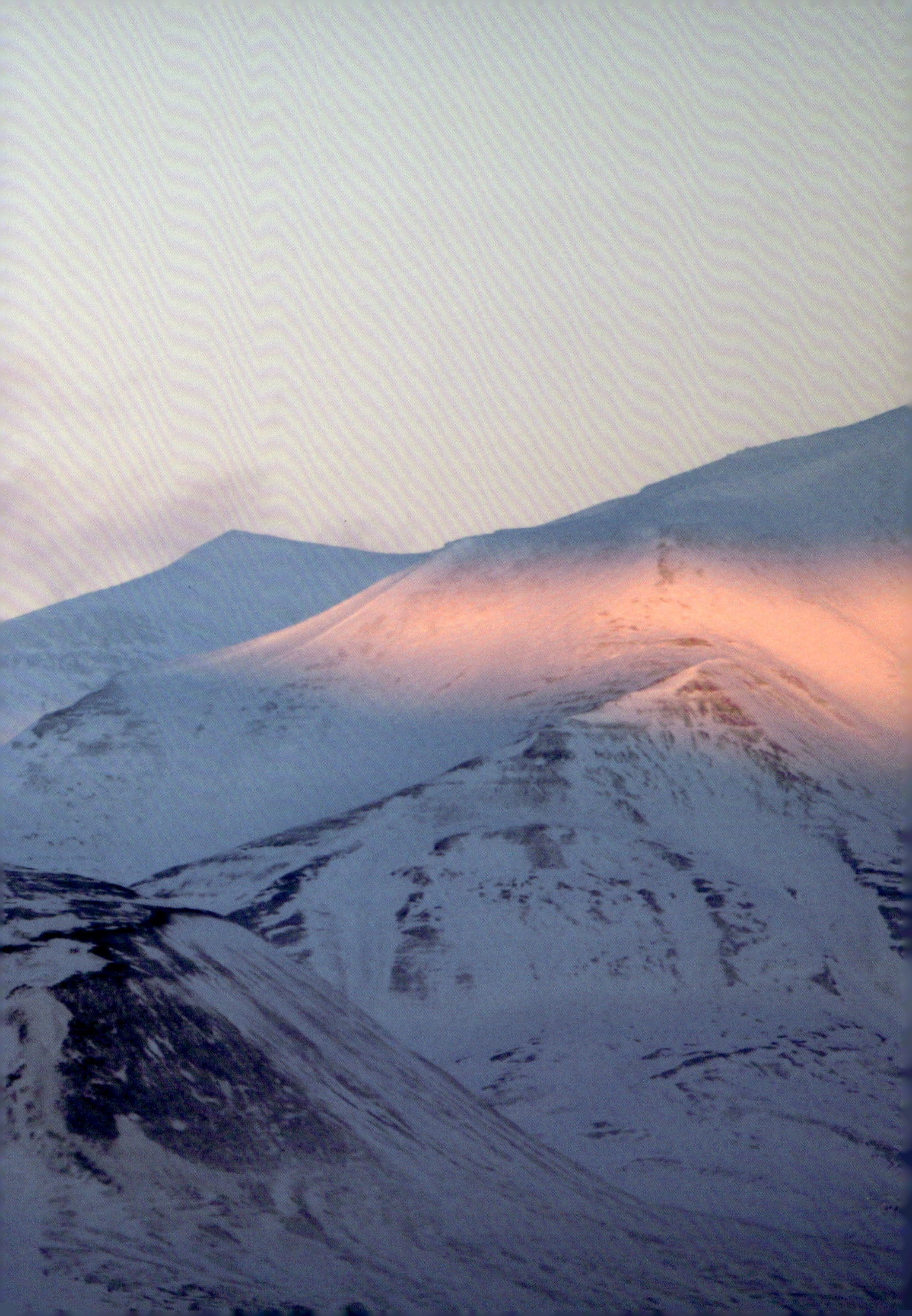

HOLI PURNIMA

Holi Purnima (*Purnima* means "full moon" in Sanskrit) is a Hindu festival celebrating the arrival of spring. It originated on the Indian subcontinent but has spread with the Indian diaspora to other countries in South Asia and elsewhere, including the Caribbean, United Kingdom, and North America.

The festival of Holi begins on the evening of the full moon in the month of Phalgun on the Hindu calendar, and lasts throughout the following day. On the first night, celebrants observe this festival of colors honoring the Hindu god Vishnu by performing religious rituals around bonfires. Throughout the next day, they shower each other with colored powders, dance in the street, and partake in Holi delicacies and ceremonial drinks.

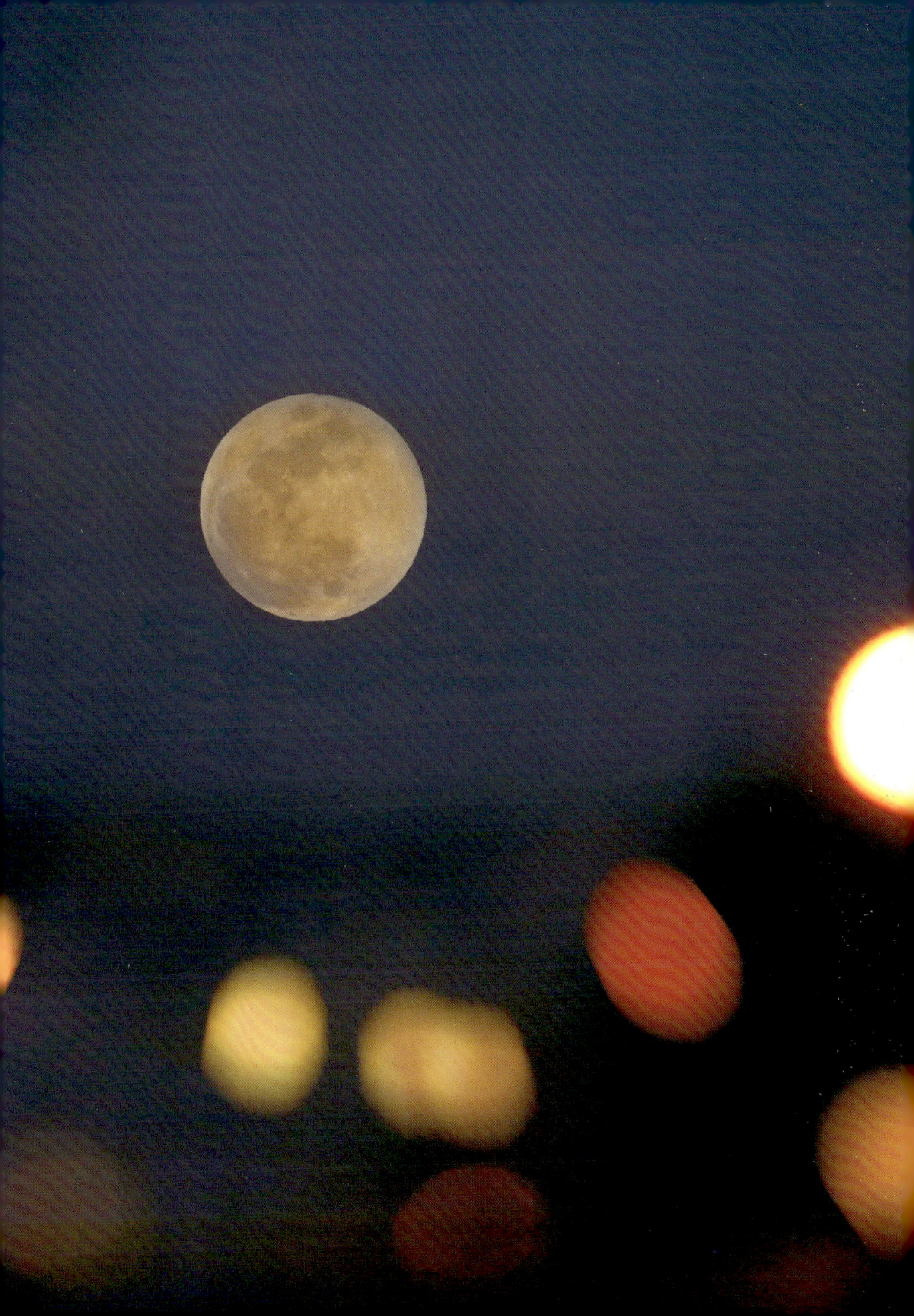

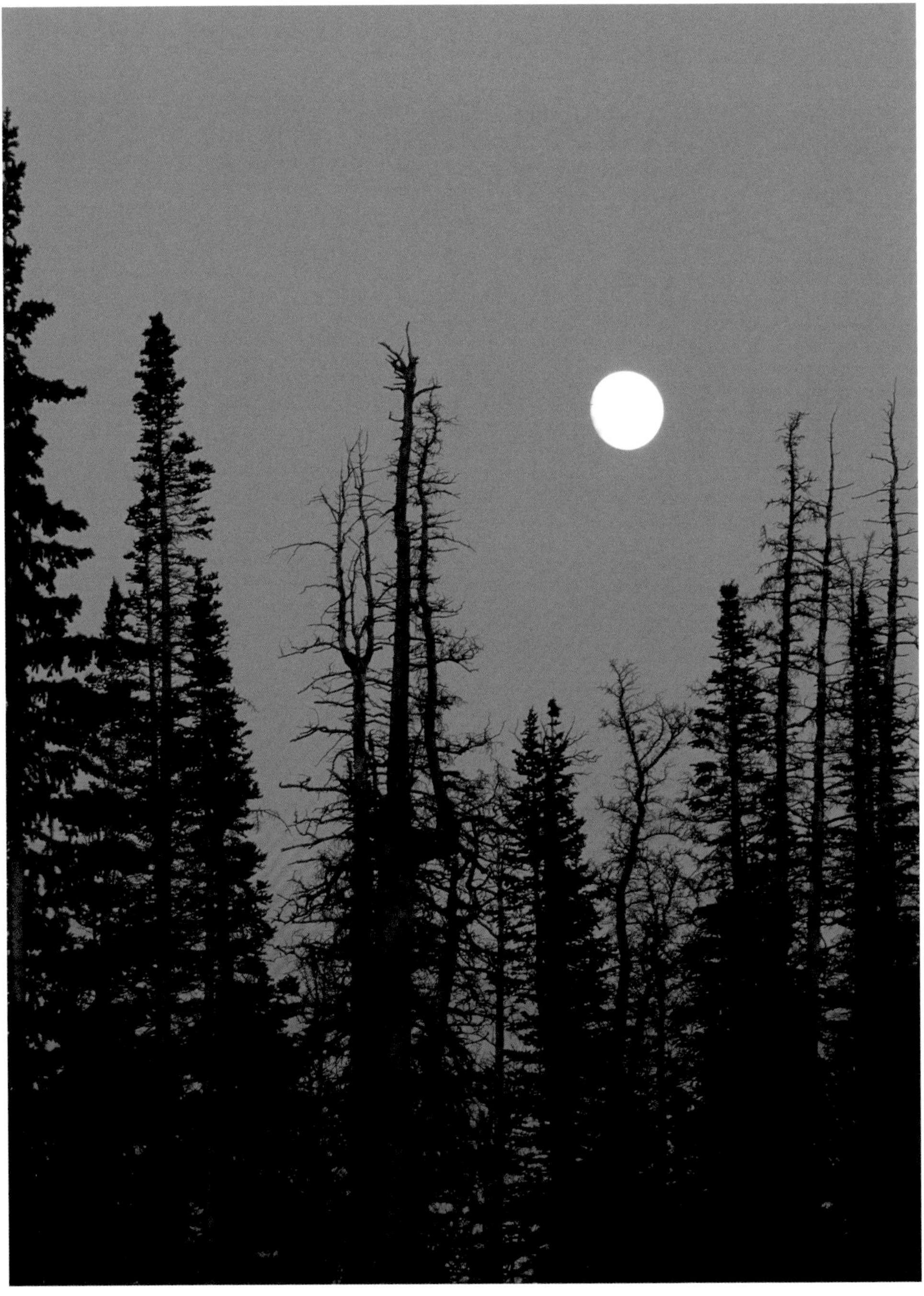

SEED MOON

April's full moon was named the Seed Moon, both by the medieval English and by some Native American tribes, as the thawed ground and the appearance of wild flowers brings an opportune time to sow seeds.

THE MOON was but a chin of gold
A night or two ago,
And now she turns her perfect face
Upon the world below.
EMILY DICKINSON

PINK MOON

The Pink Moon derives its name not from the moon's color but from the seasonal flowering of moss pink, or perennial wild ground phlox. The bright pink flowers are native to the eastern United States, and their appearance was typically associated with the arrival of spring in the region.

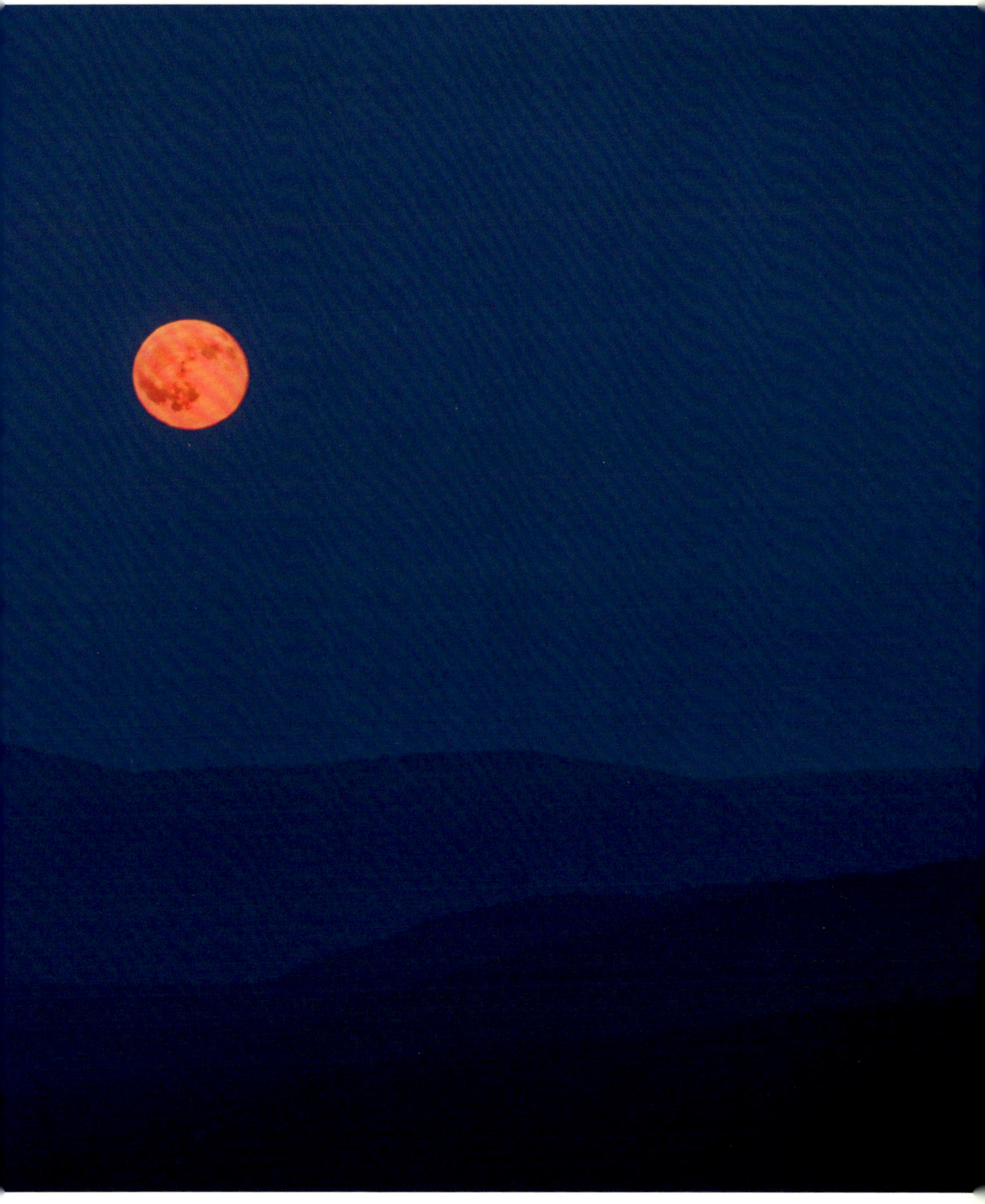

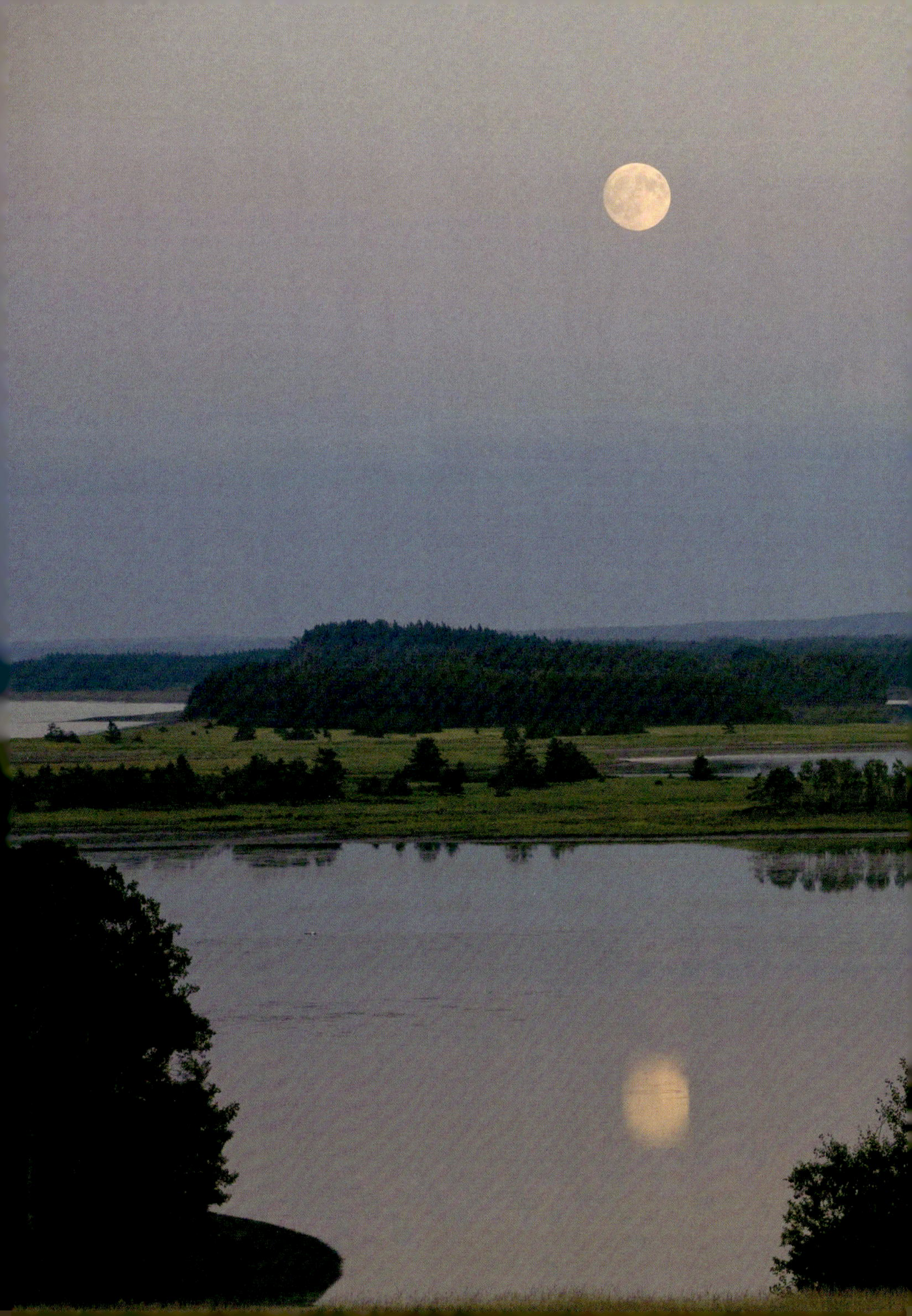

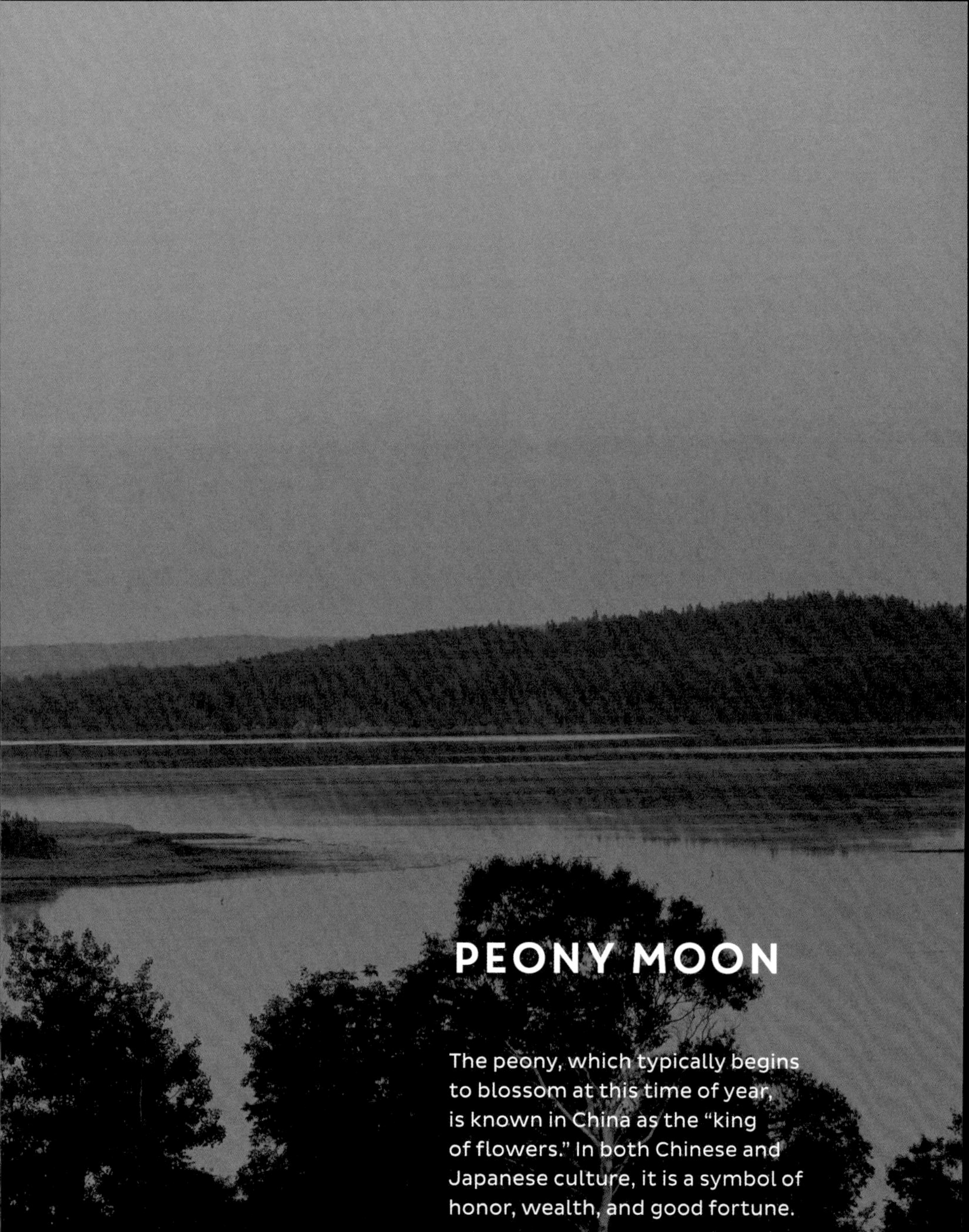

PEONY MOON

The peony, which typically begins to blossom at this time of year, is known in China as the "king of flowers." In both Chinese and Japanese culture, it is a symbol of honor, wealth, and good fortune.

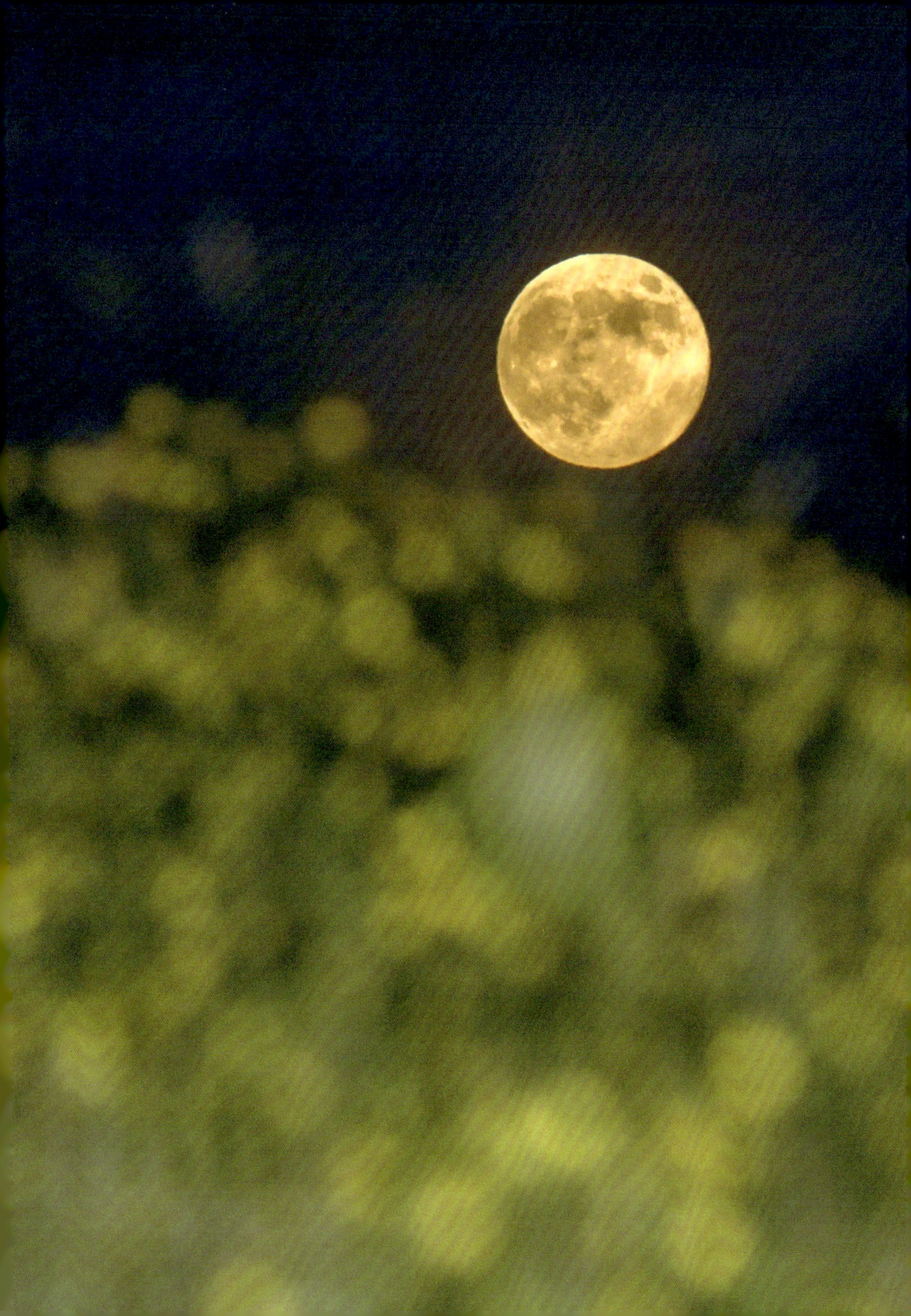

FLOWER MOON

By May, snows have disappeared, ice has melted, and seeds sown in April have begun to sprout. Flowers, wild and cultivated, start to bloom, giving May's full moon the name Flower Moon.

The moon, like a flower
In heaven's high bower,
With silent design
Sits and smiles on the night.
WILLIAM BLAKE

DRAGON MOON

In Chinese mythology, when the heavenly dragon constellation was at its zenith, the moon would shine with magical light. And so it became known as the Dragon Moon. In the light of the Dragon Moon, it was said the old would become young; the infirm, healthy; the weary, reinvigorated. Such was the power of the moon that the Chinese called it "the pearl of heaven."

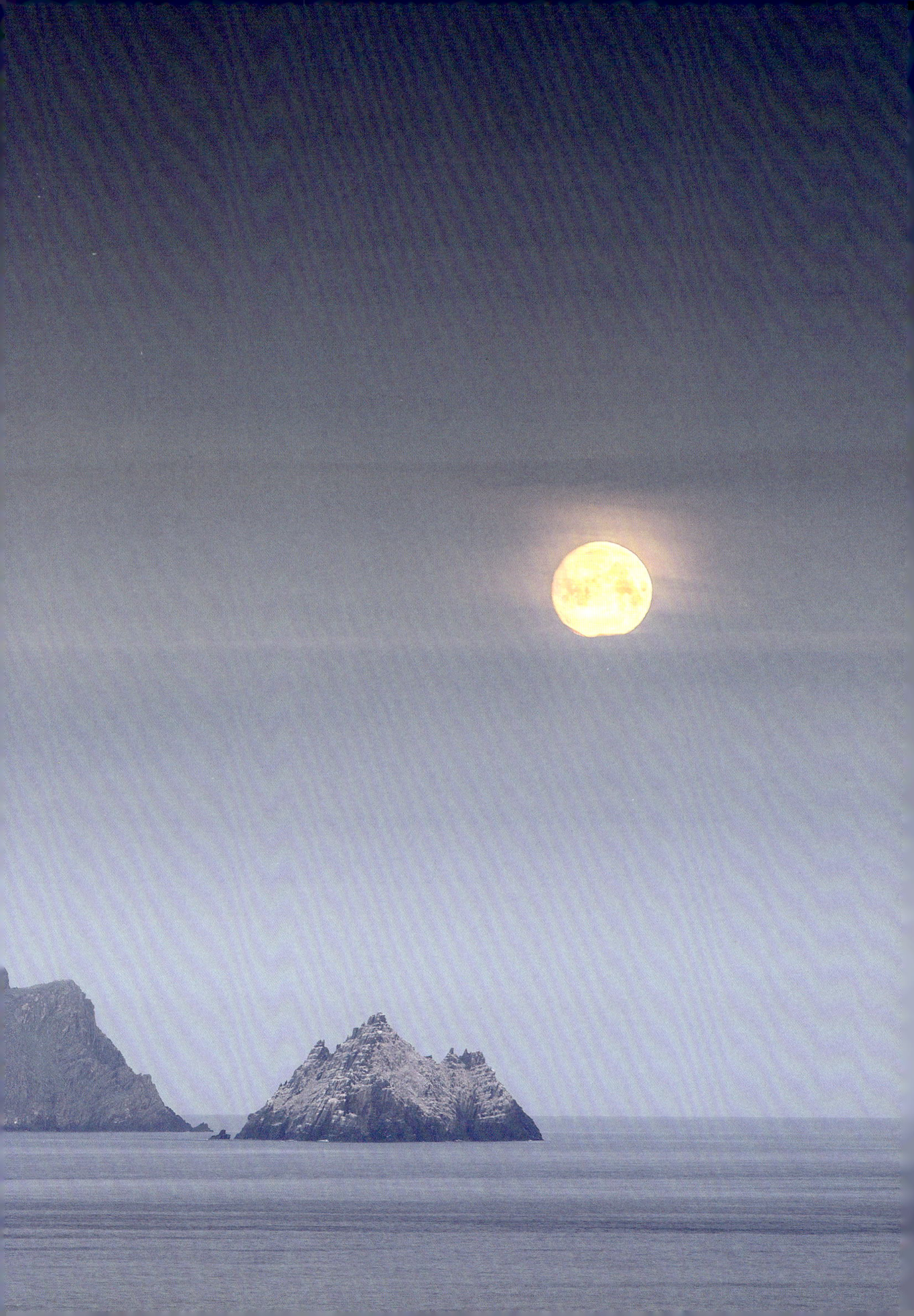

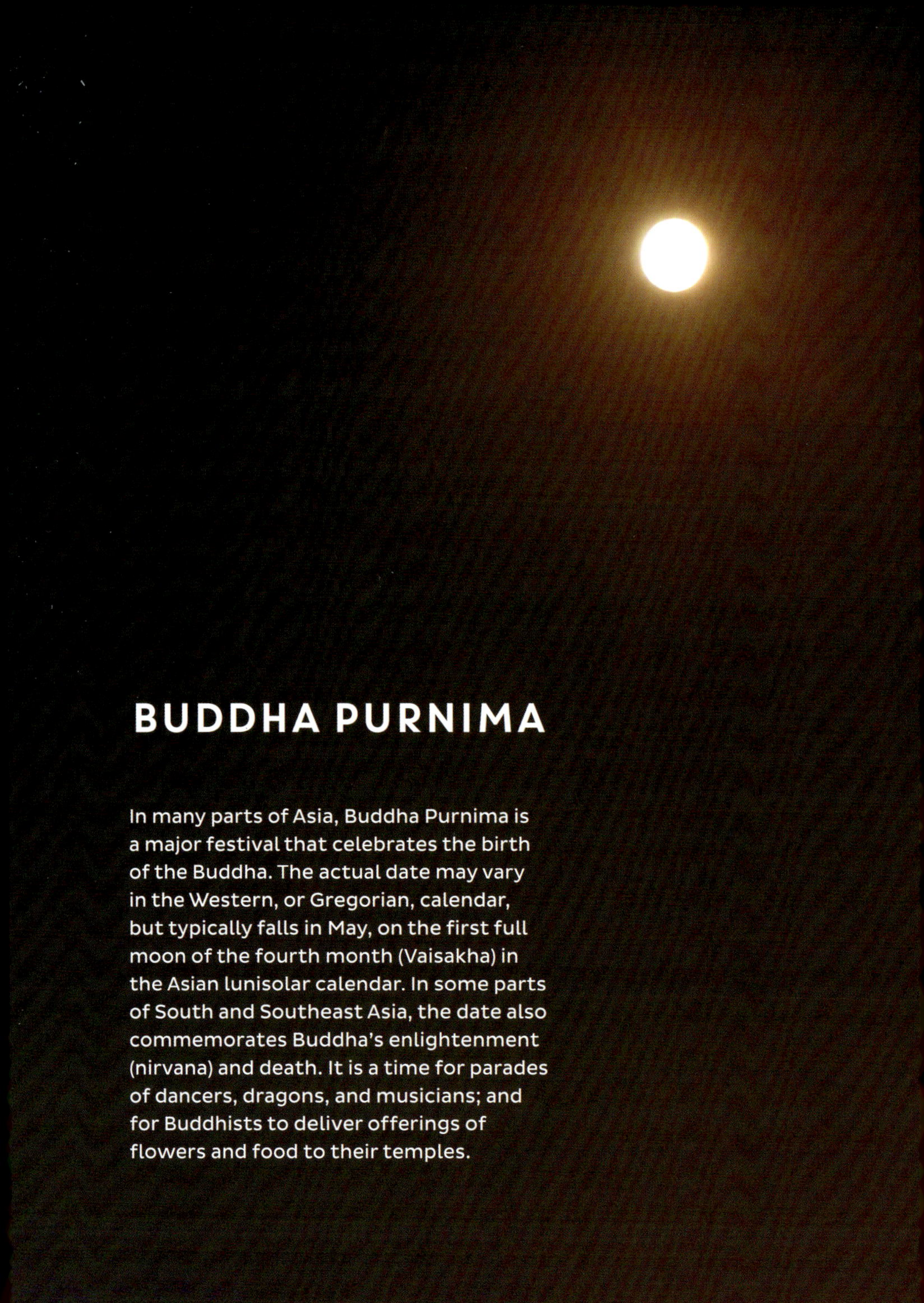

BUDDHA PURNIMA

In many parts of Asia, Buddha Purnima is a major festival that celebrates the birth of the Buddha. The actual date may vary in the Western, or Gregorian, calendar, but typically falls in May, on the first full moon of the fourth month (Vaisakha) in the Asian lunisolar calendar. In some parts of South and Southeast Asia, the date also commemorates Buddha's enlightenment (nirvana) and death. It is a time for parades of dancers, dragons, and musicians; and for Buddhists to deliver offerings of flowers and food to their temples.

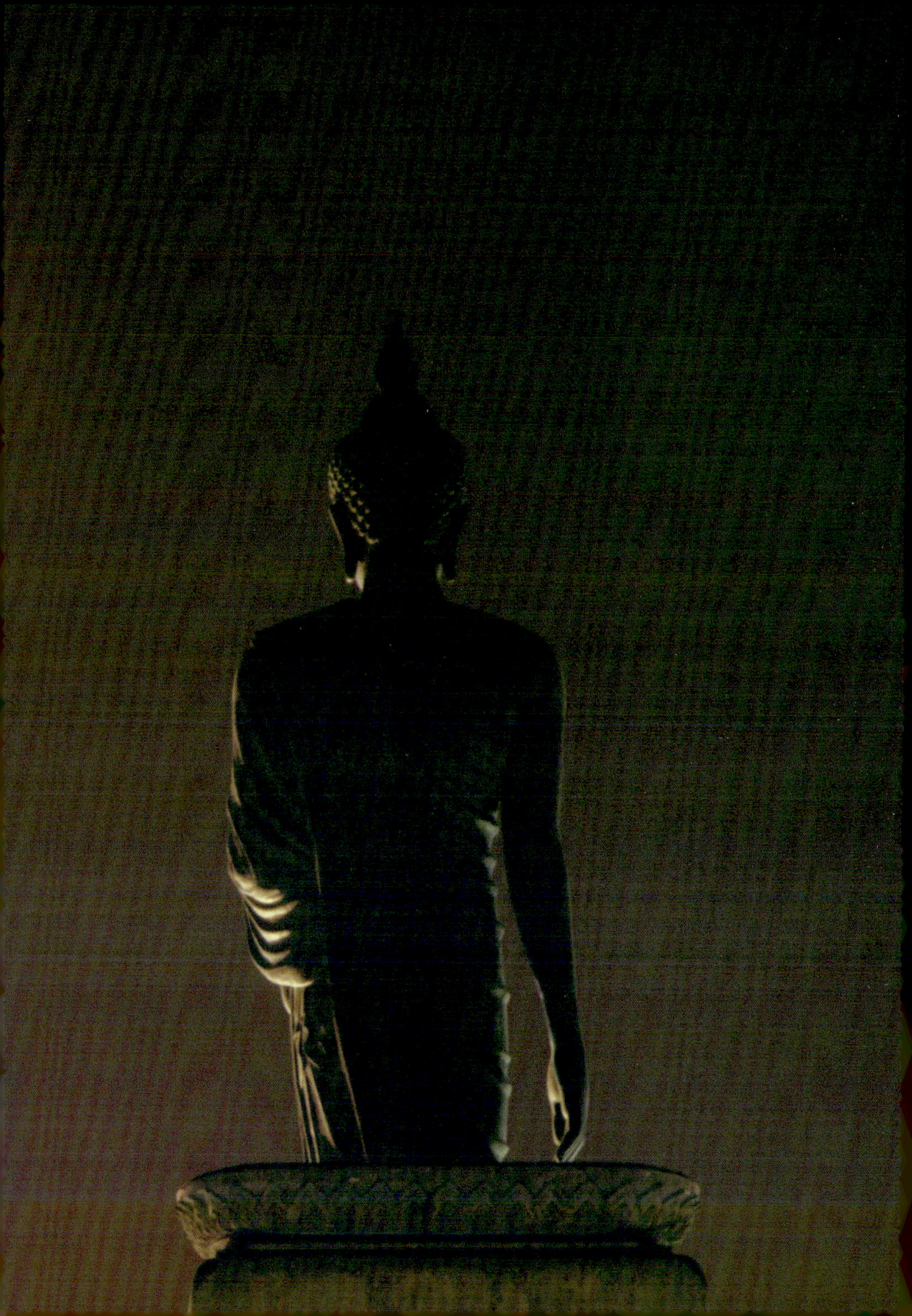

STRAWBERRY MOON

In North America, the Algonquian tribes (among others) called June's full moon the Strawberry Moon because it marked the time of year when the wild strawberries were ripe for picking. In Europe it was called the Rose Moon, for the season when roses are in bloom.

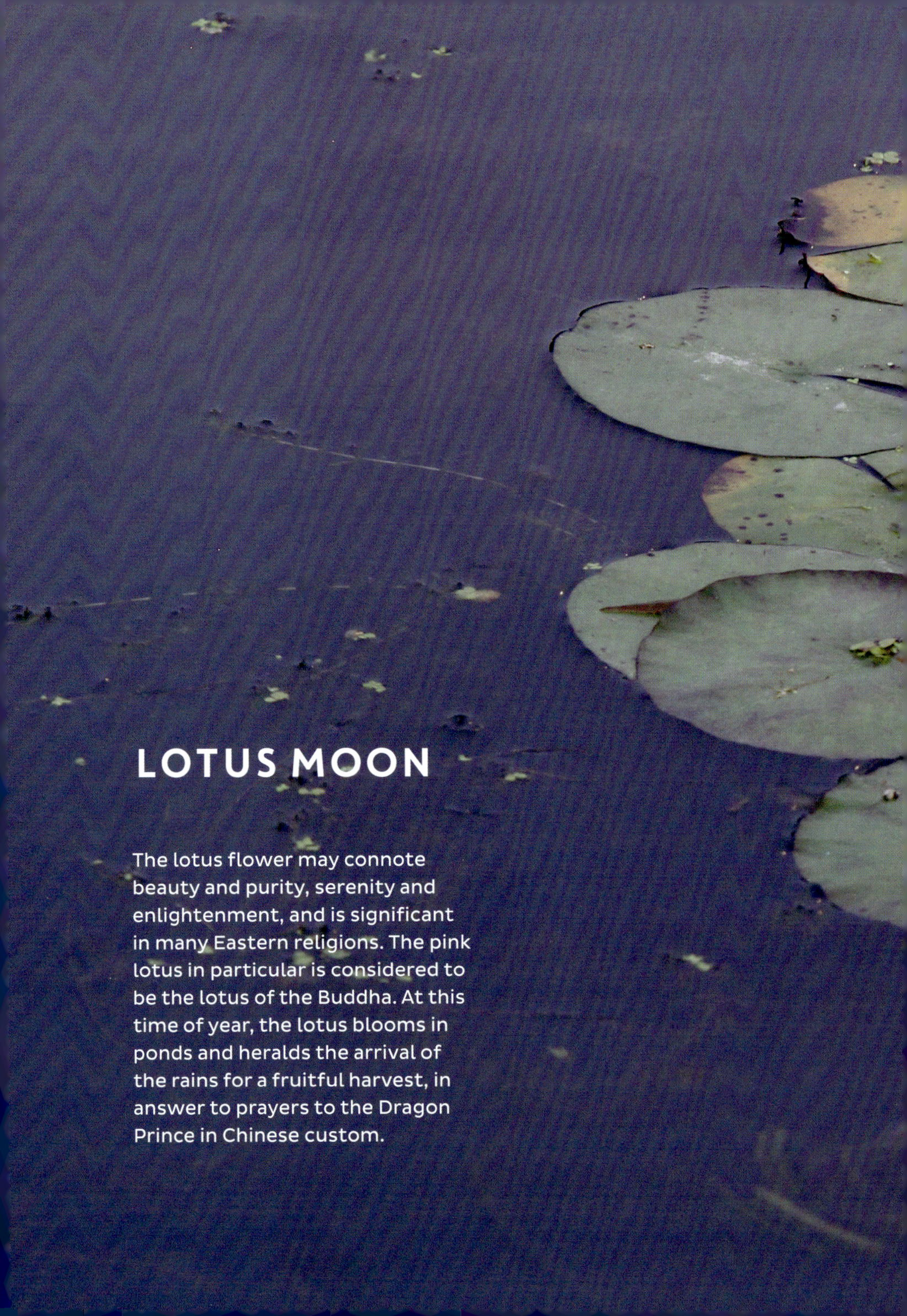

LOTUS MOON

The lotus flower may connote beauty and purity, serenity and enlightenment, and is significant in many Eastern religions. The pink lotus in particular is considered to be the lotus of the Buddha. At this time of year, the lotus blooms in ponds and heralds the arrival of the rains for a fruitful harvest, in answer to prayers to the Dragon Prince in Chinese custom.

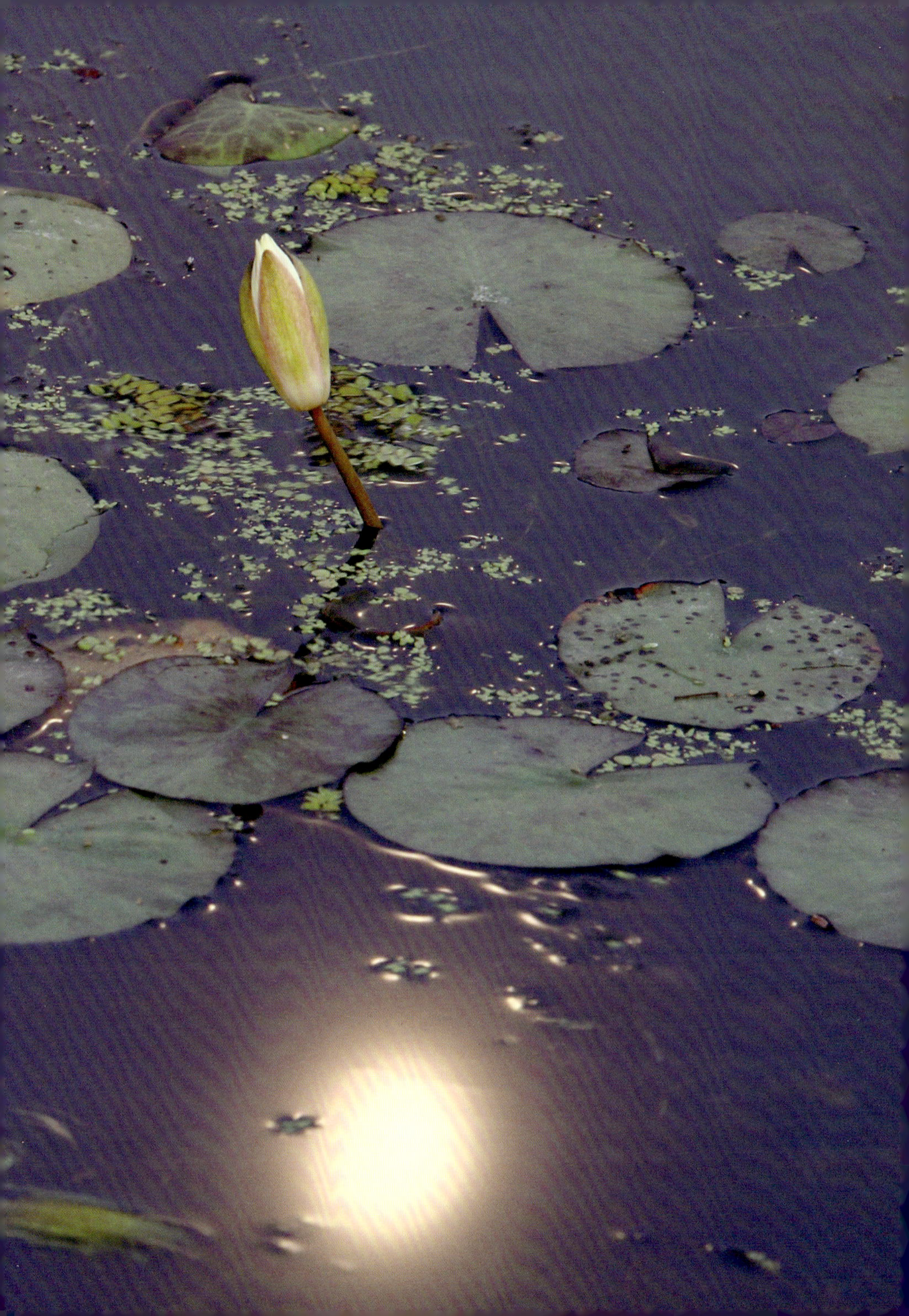

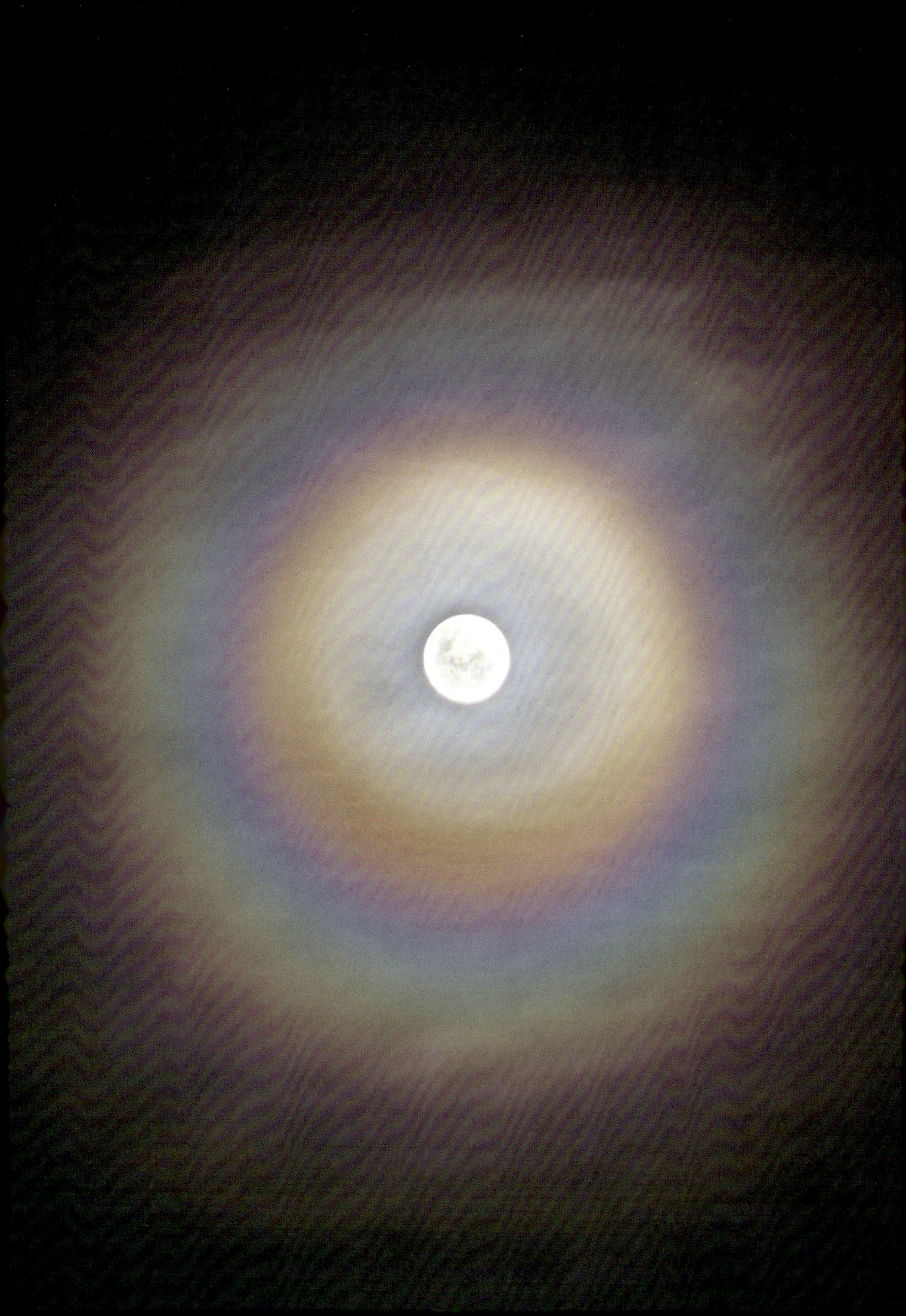

DYAD MOON

In some pagan traditions, June's full moon was referred to as the Dyad Moon, from the Latin for "pair" or "two." According to one legend, it symbolized the human and the divine, or the goddess and the god. Falling as it does midway through the year near the Summer Solstice, the longest day of the year, it may also represent the partnership between the Sun and the Moon.

They danced by the light of the moon,
The moon,
The moon,
They danced by the light of the moon.
EDWARD LEAR

MEAD MOON

The full Mead Moon is named for the time when the first crops of the year were being harvested, and medieval English farmers could enjoy the rewards of their labors. This was an occasion to celebrate with a drink of mead, a fermented beverage made from honey that was called the "nectar of the gods," and is one of the oldest alcoholic drinks in recorded history.

With freedom, books, flowers, and the moon who could not be happy? OSCAR WILDE

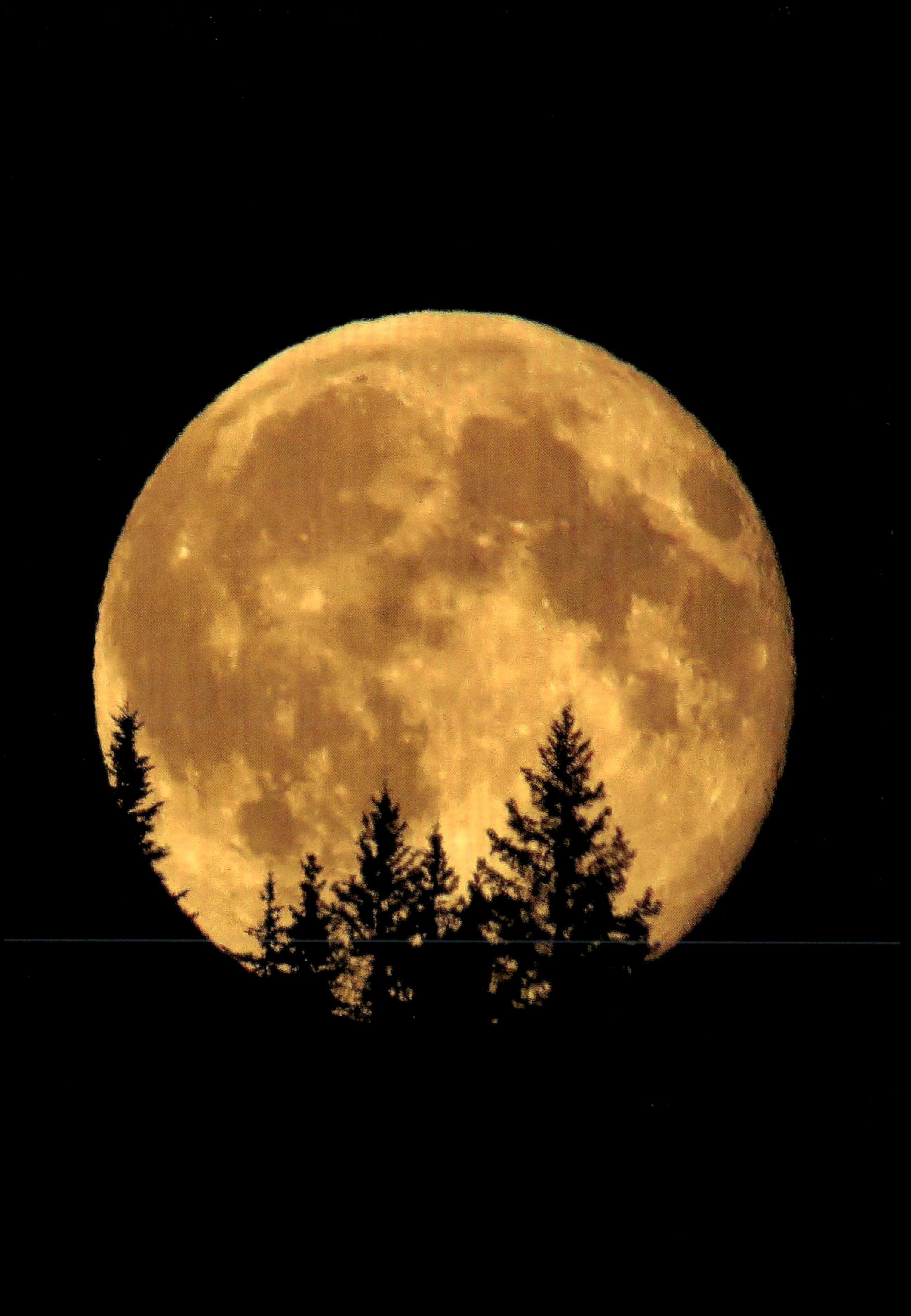

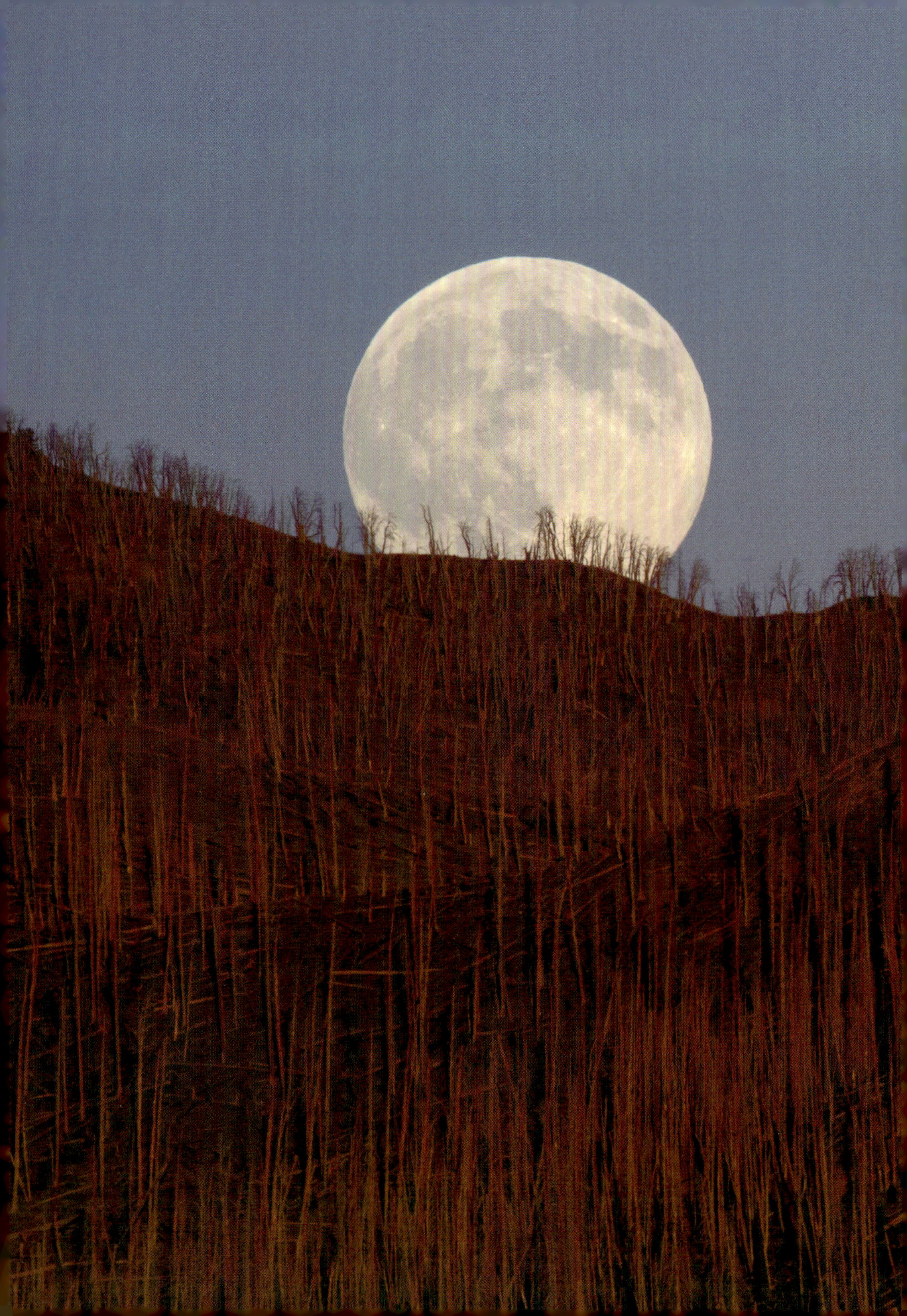

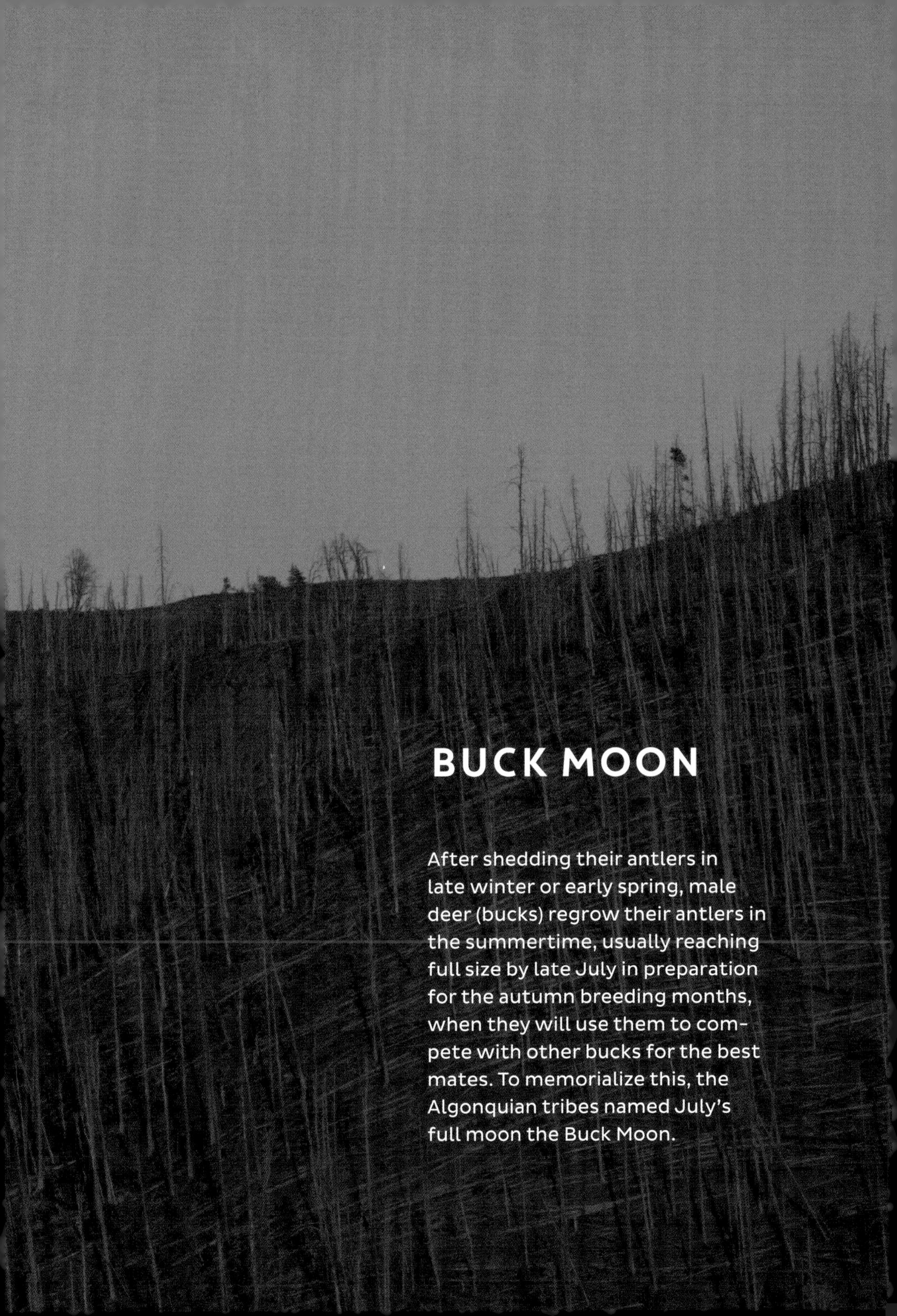

BUCK MOON

After shedding their antlers in late winter or early spring, male deer (bucks) regrow their antlers in the summertime, usually reaching full size by late July in preparation for the autumn breeding months, when they will use them to compete with other bucks for the best mates. To memorialize this, the Algonquian tribes named July's full moon the Buck Moon.

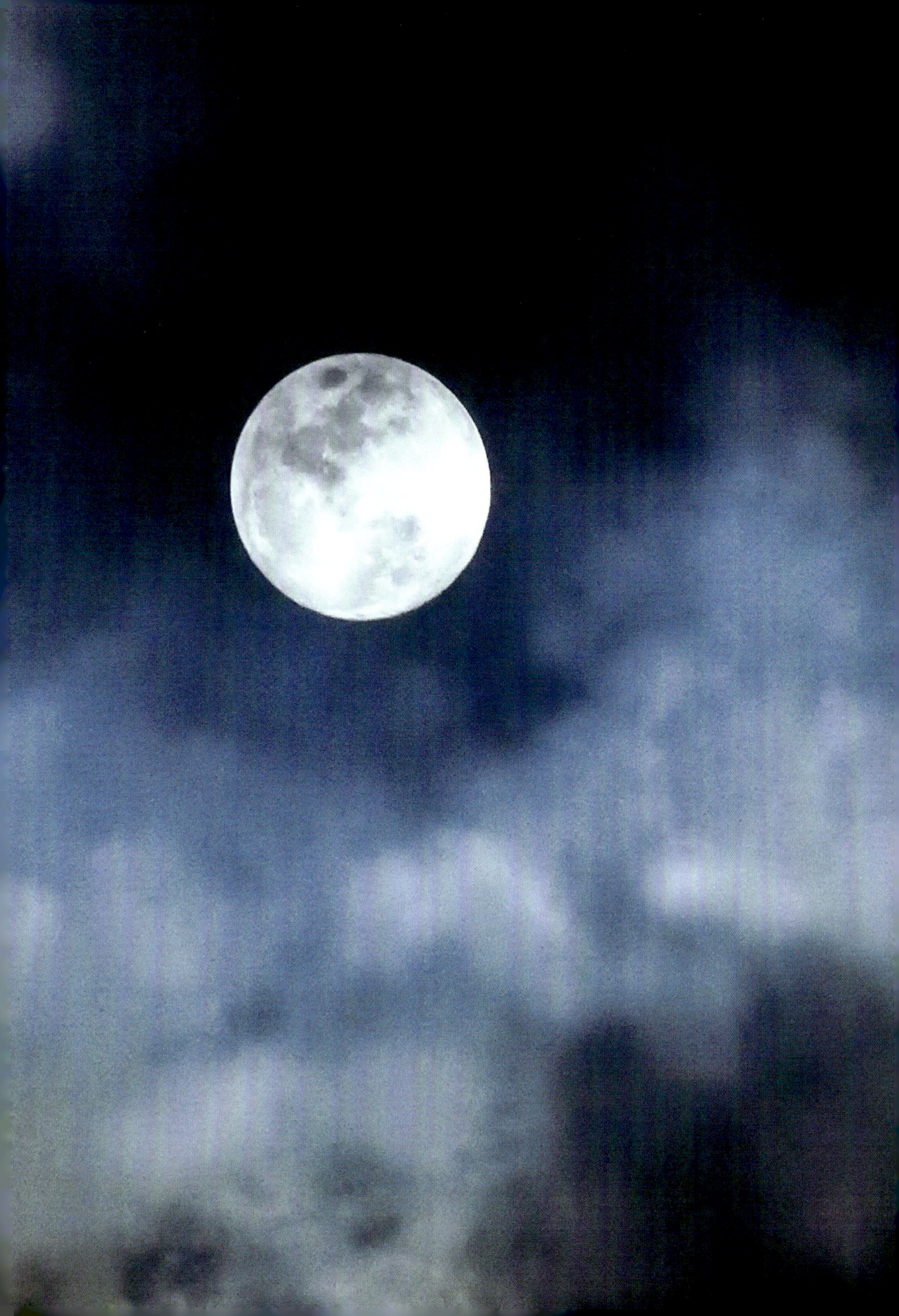

HUNGRY GHOST MOON

The Hungry Ghost Moon is named for the belief in Chinese Taoism and Buddhism that spirits from the other world roam freely in the seventh month of the lunar calendar (July or August in the Western calendar). The Ghost Festival is held during this full moon, when families pray to their deceased relatives, hold feasts, burn joss notes, and leave offerings to appease wandering spirits and ward off bad luck.

RED MOON

The August full moon is sometimes known as the Red Moon; when the moon is rising or setting low in the late summer sky, the hazy atmosphere can cast the moon in varying shades of red. In some coastal regions, summer fog can also produce a similar reddening effect. This is not the same as the reddish moon caused by a lunar eclipse, more commonly known as a Blood Moon.

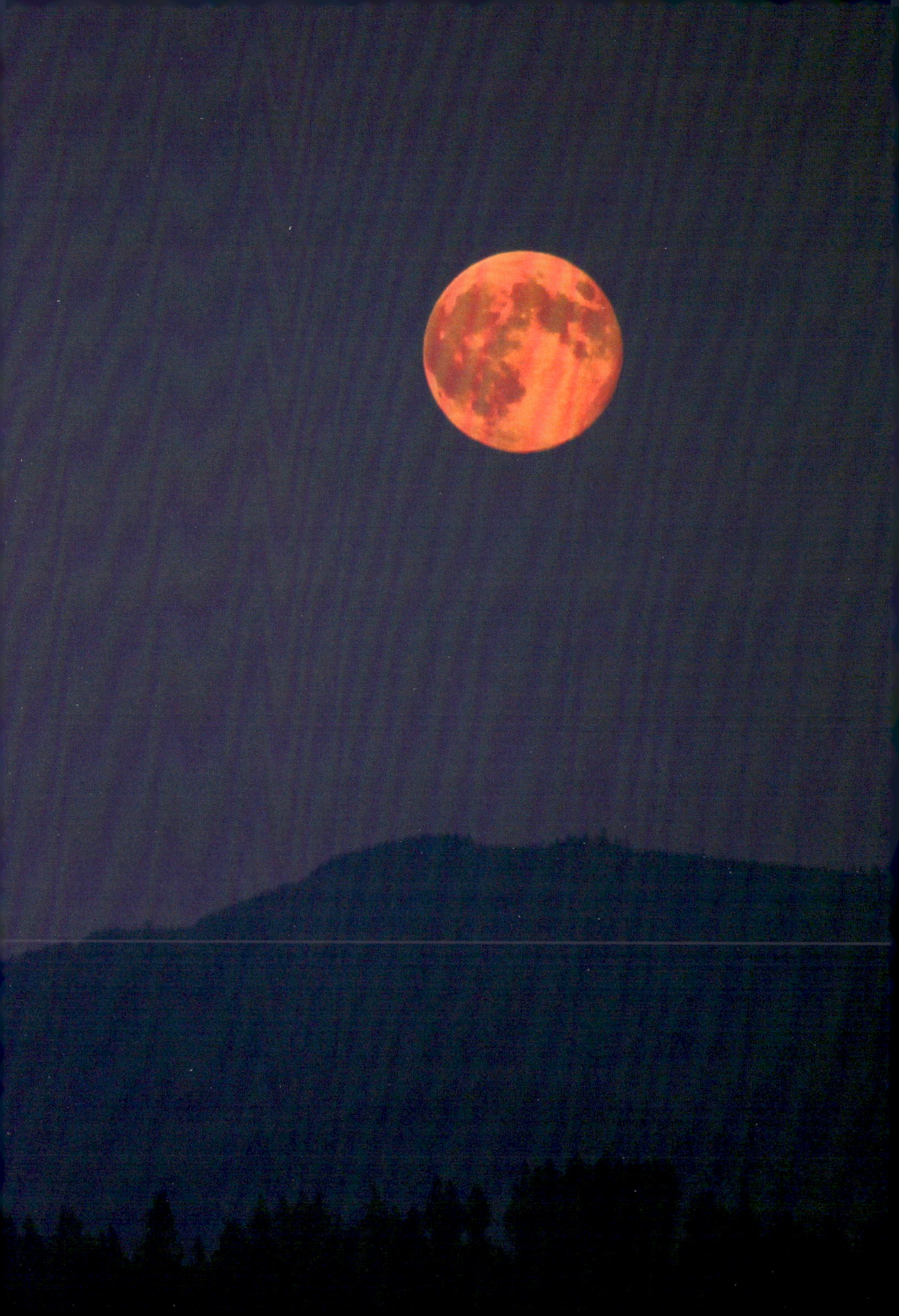

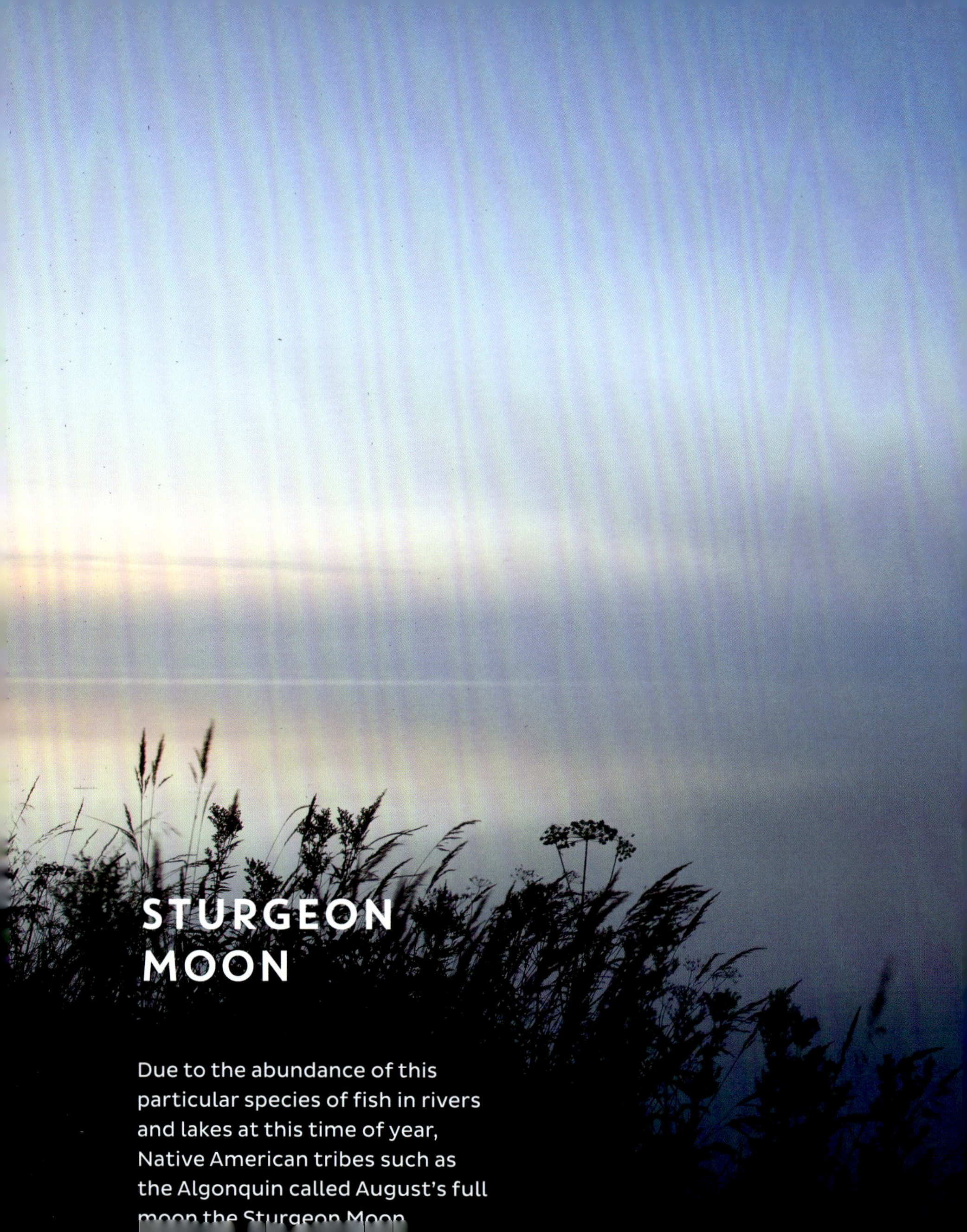

STURGEON MOON

Due to the abundance of this particular species of fish in rivers and lakes at this time of year, Native American tribes such as the Algonquin called August's full moon the Sturgeon Moon.

SWAN FLIGHT MOON

In Inuit custom, the August full moon was named in observance of this time of year when young swans take their first flight.

The moon lives in the lining of your skin.
PABLO NERUDA

HARVEST MOON

The full moon closest to September's autumnal equinox, and the start of fall, is the Northern Hemisphere's familiarly named Harvest Moon. This moon is particularly significant as the earliest moonrise of the year, with no interval of time between sunset and moonrise. In a less-mechanized era, the bright moonlight that accompanies the Harvest Moon brought a welcome light to farmers harvesting their grain crops past sunset, and was also known as the Corn or Barley Moon.

So, we'll go no more A-roving
So late into the night,
Though the heart be still as loving
And the moon be still bright.
LORD BYRON

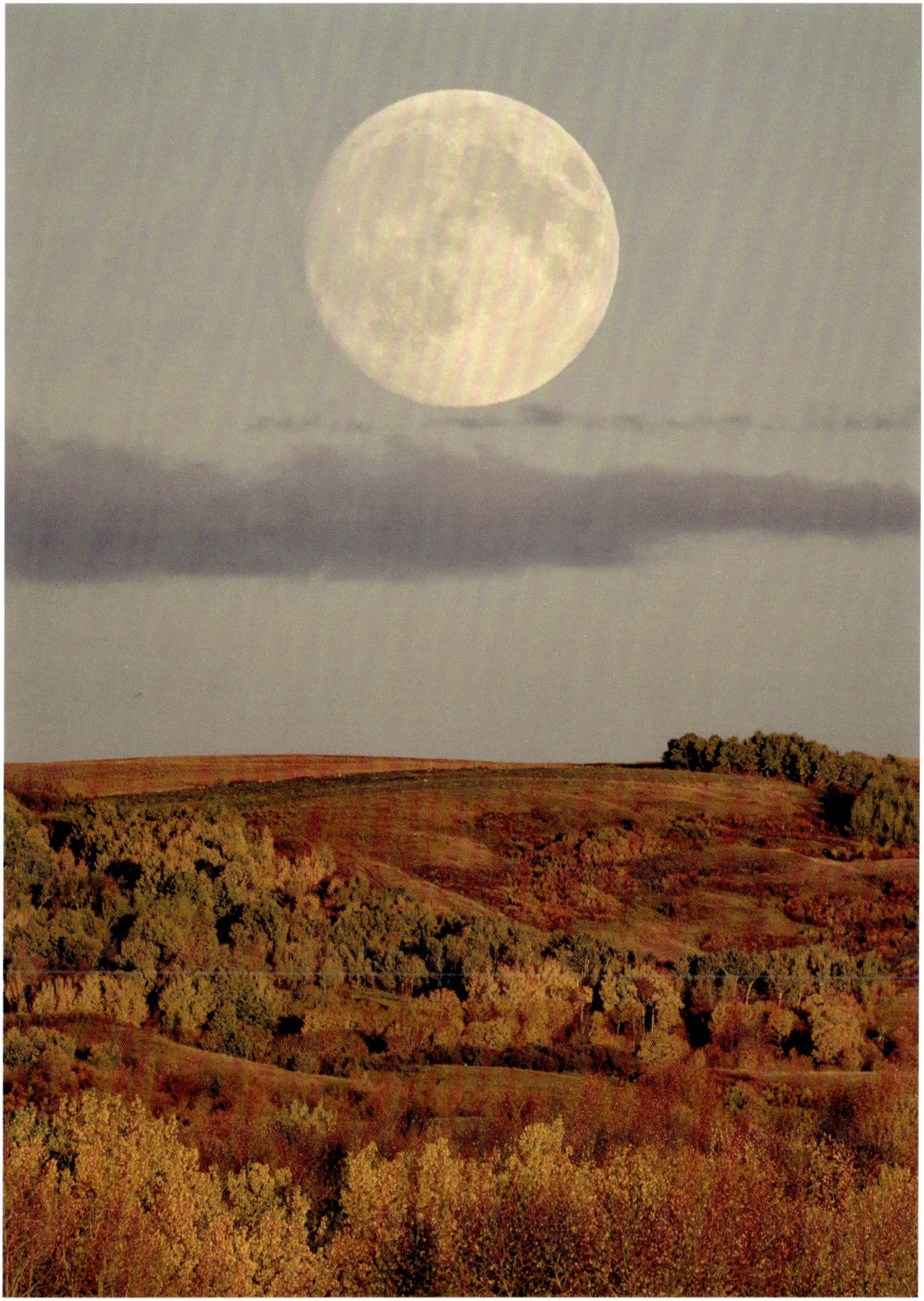

CHUSHU MOON

On the traditional Japanese or lunisolar calendar, the autumn full moon festival—Tsukimi—fell on the fifteenth day of the eighth month to celebrate the fall harvest. This full moon is called the Chushu Moon ("Mid-Autumn" Moon) and ceremonial viewings of this moon in Japan date back over a thousand years, when the practice was brought over from China.

Tsukimi ("moon viewing") is celebrated across Japan at about the time of the autumnal equinox, with viewing parties and foods such as rice dumplings (Tsukimi dango) that resemble the full moon, soba and udon noodles topped with a raw egg (which is symbolic of the full moon), and sake. Other Tsukimi traditions have included decorating the home with Japanese pampas grass (susuki), and reciting poetry under the moonlight.

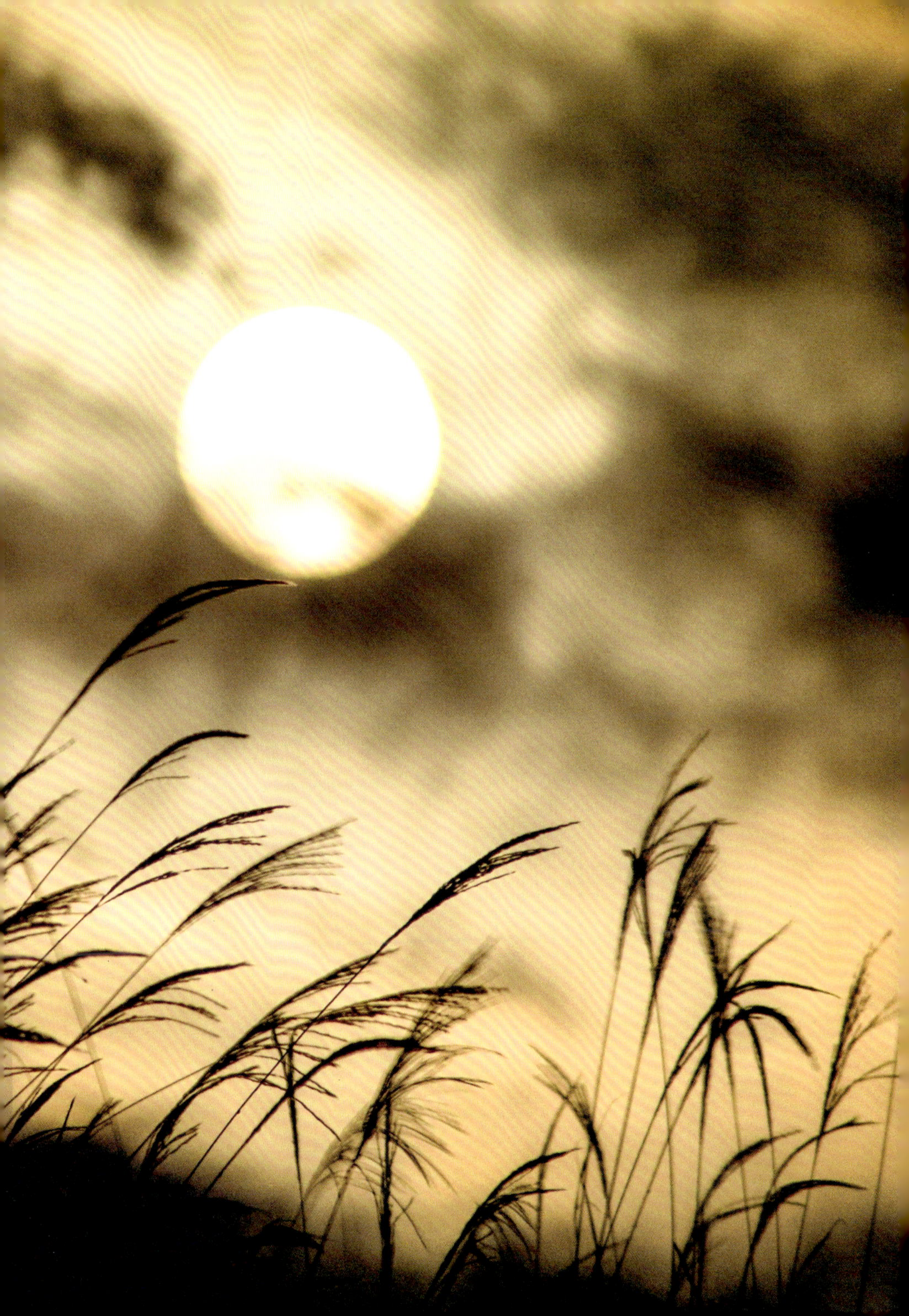

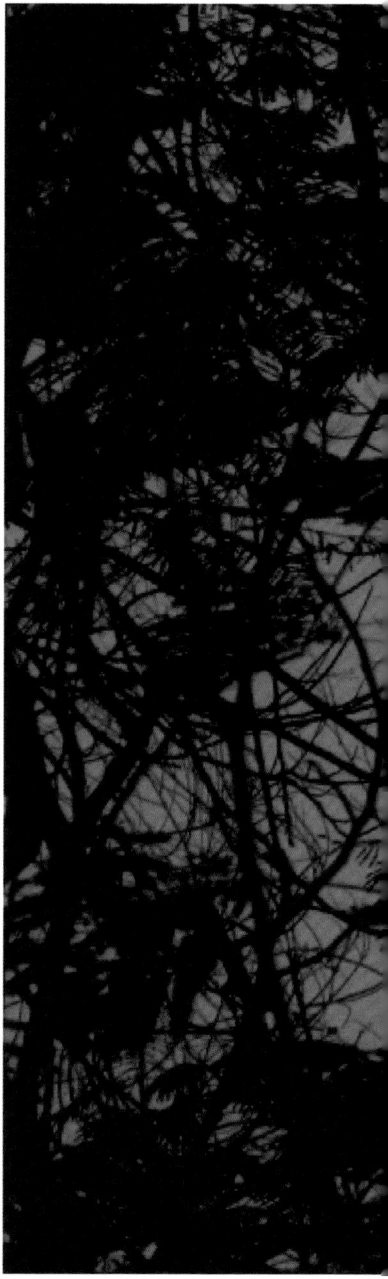

BHADRAPADA PURNIMA

Bhadrapada Purnima falls in the sixth month of the Hindu calendar when the night of the full moon carries religious significance. At this time, devotees pay homage to Lord Vishnu, beginning with a ritual bath in rivers such as the Yamuna or Ganges, or other waters imbued with sanctity. While it is a day of fasting, the Bhadrapada Purnima is also marked by festivals and celebration.

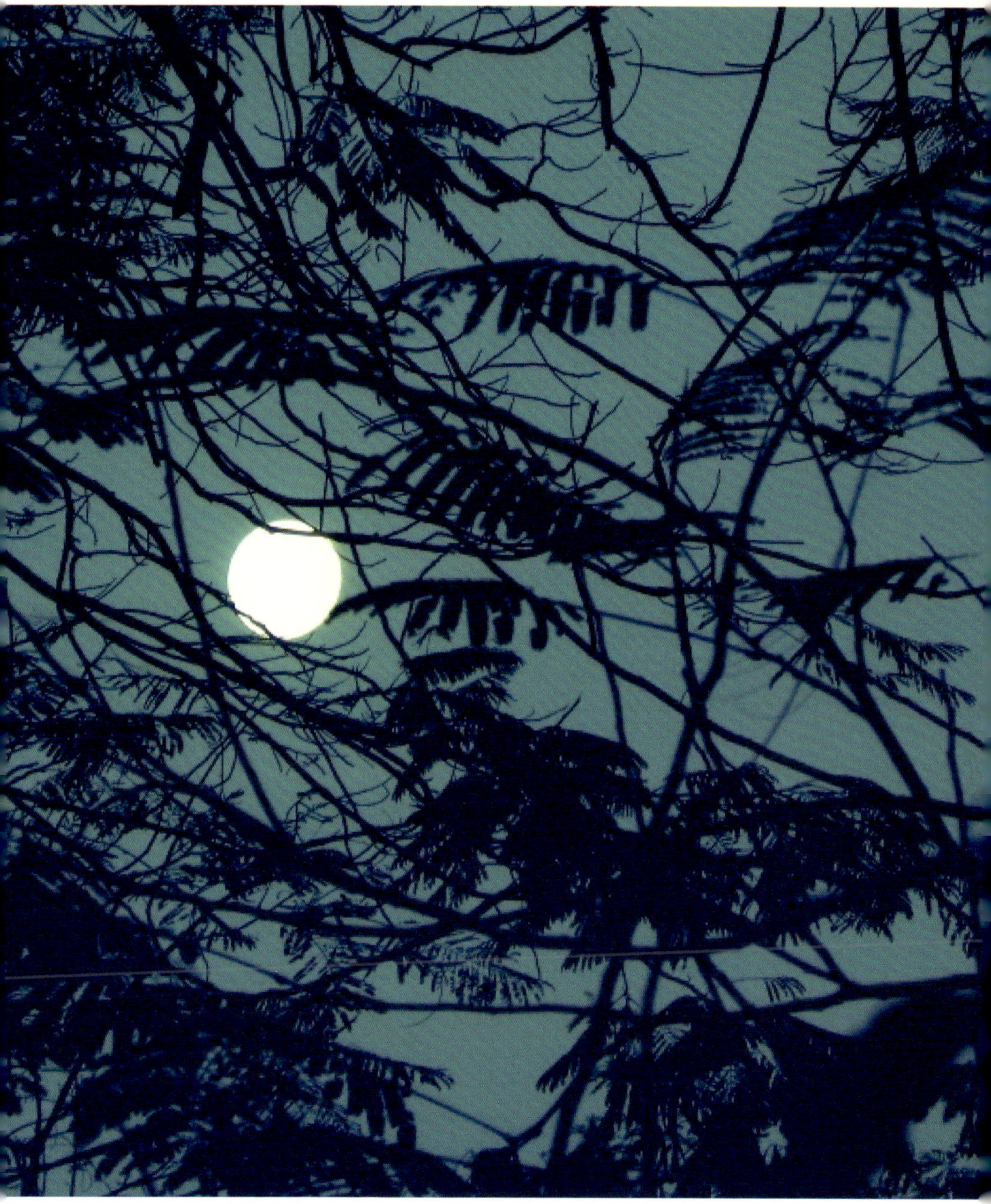

HUNTER'S MOON

The Hunter's Moon is the first full moon following the Harvest Moon in October (except once every four years when it appears in November). During this season, prey animals are often well-fattened in preparation for the winter months, and have fewer places to hide in the harvested fields. Accordingly, the Hunter's Moon is named in the Native American tradition as the best time to hunt and stock up on food, and is also marked by an early moonrise and a bright moon.

The Hunter's Moon waxed round in the night sky, and put to flight all the lesser stars. J.R.R. TOLKIEN

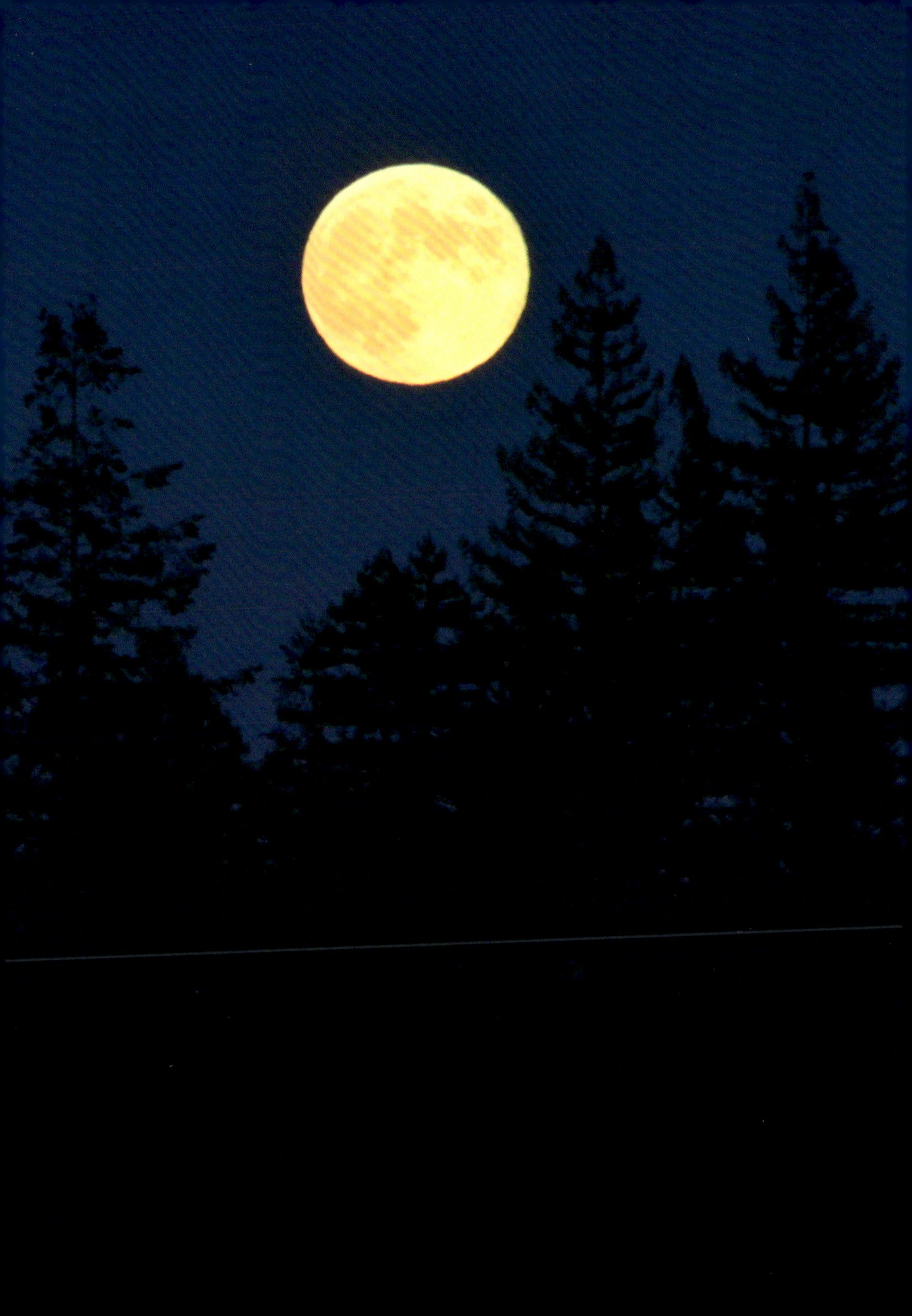

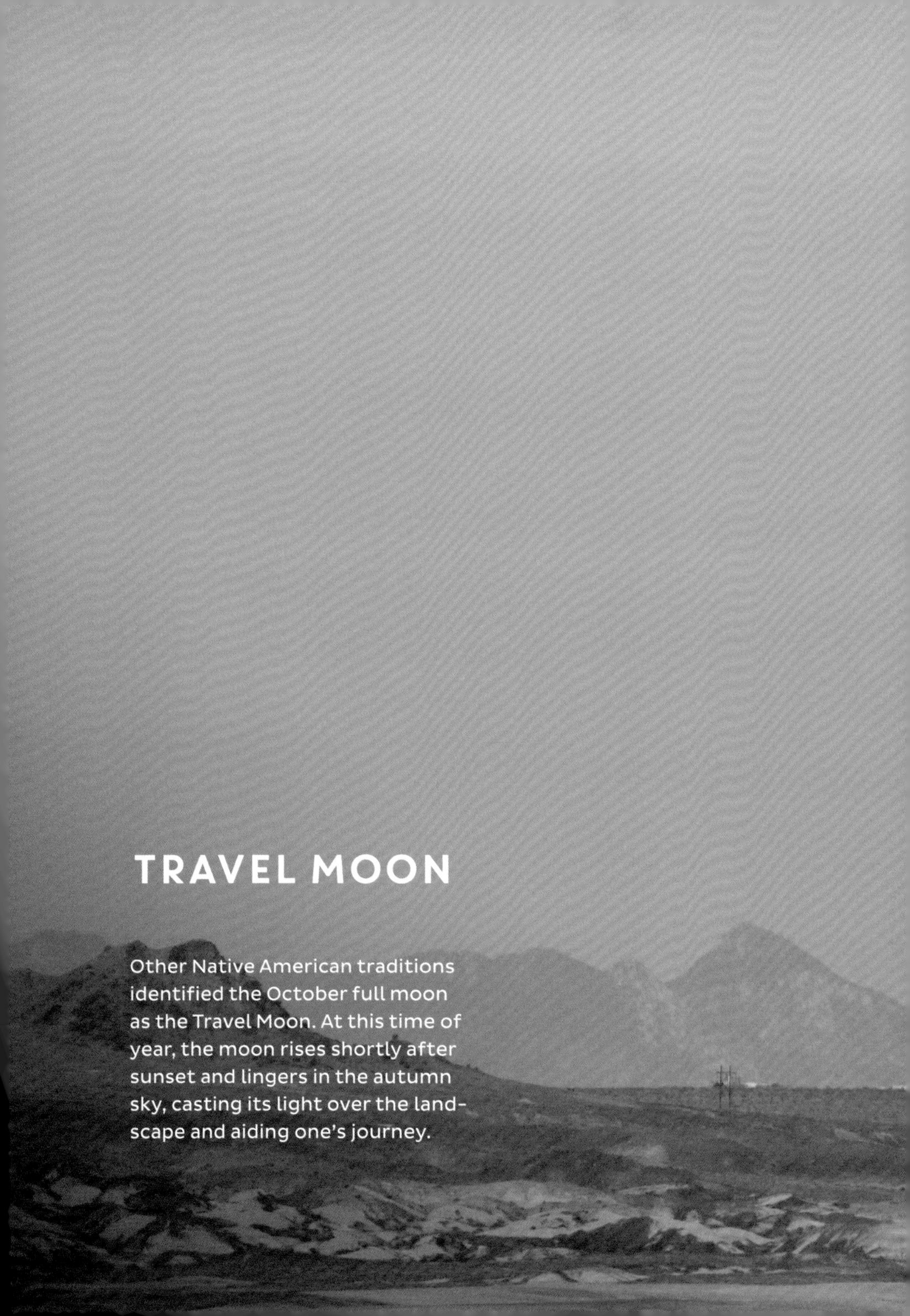

TRAVEL MOON

Other Native American traditions identified the October full moon as the Travel Moon. At this time of year, the moon rises shortly after sunset and lingers in the autumn sky, casting its light over the landscape and aiding one's journey.

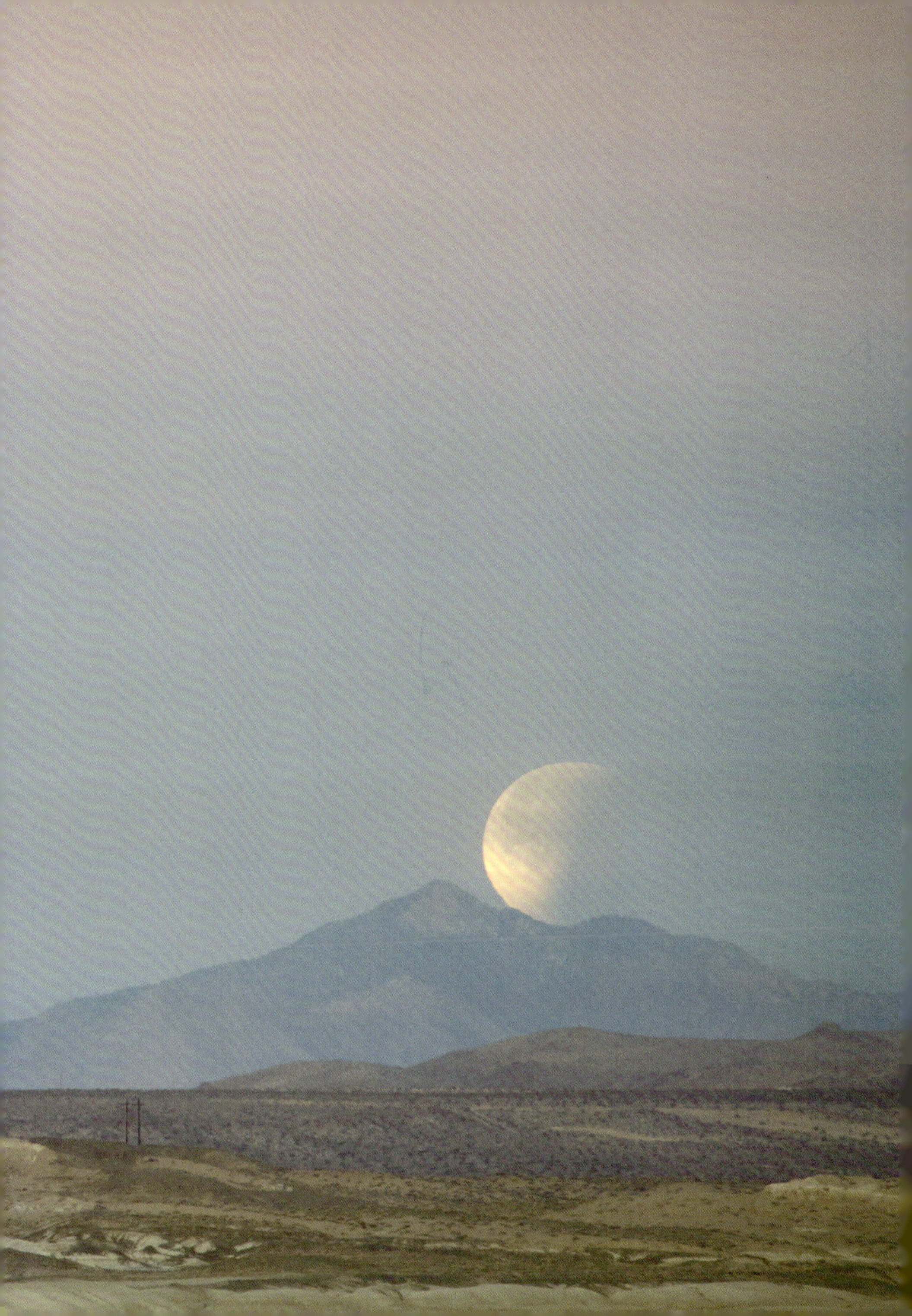

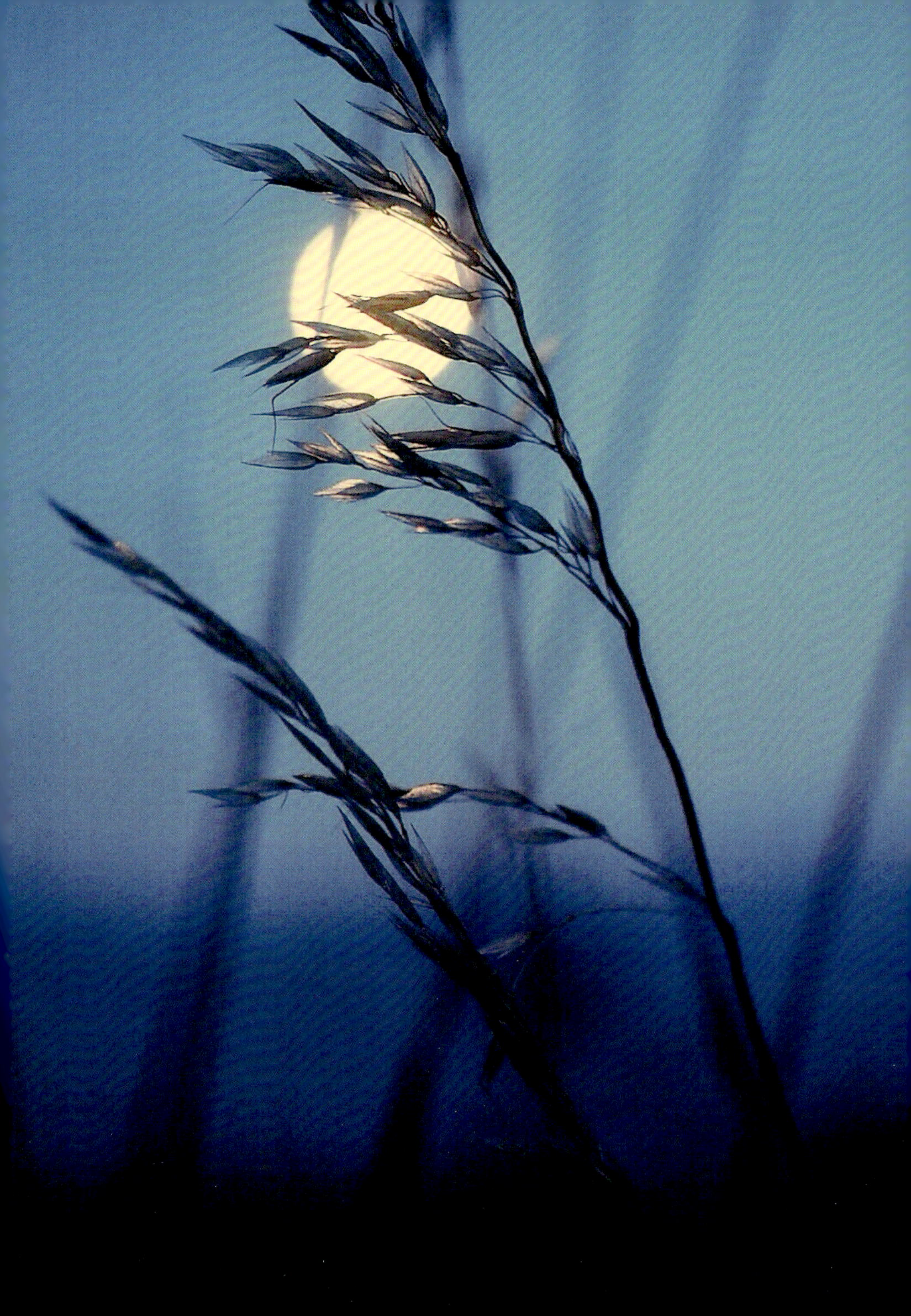

KINDLY MOON

Long before European farmers discovered techniques for planting winter crops, Chinese farmers were planting winter wheat to reap in the spring. Thus, they named October's full moon the Kindly Moon, because the first light snows of the season would melt into the warm ground, softening and hydrating it for easy planting.

FROST MOON

Some Native American tribes marked the November full moon as the Frost Moon, with the arrival of the first frosts at the end of fall, along with the likely first snows of the season.

There above them it hung, that moon.
VIRGINIA WOOLF

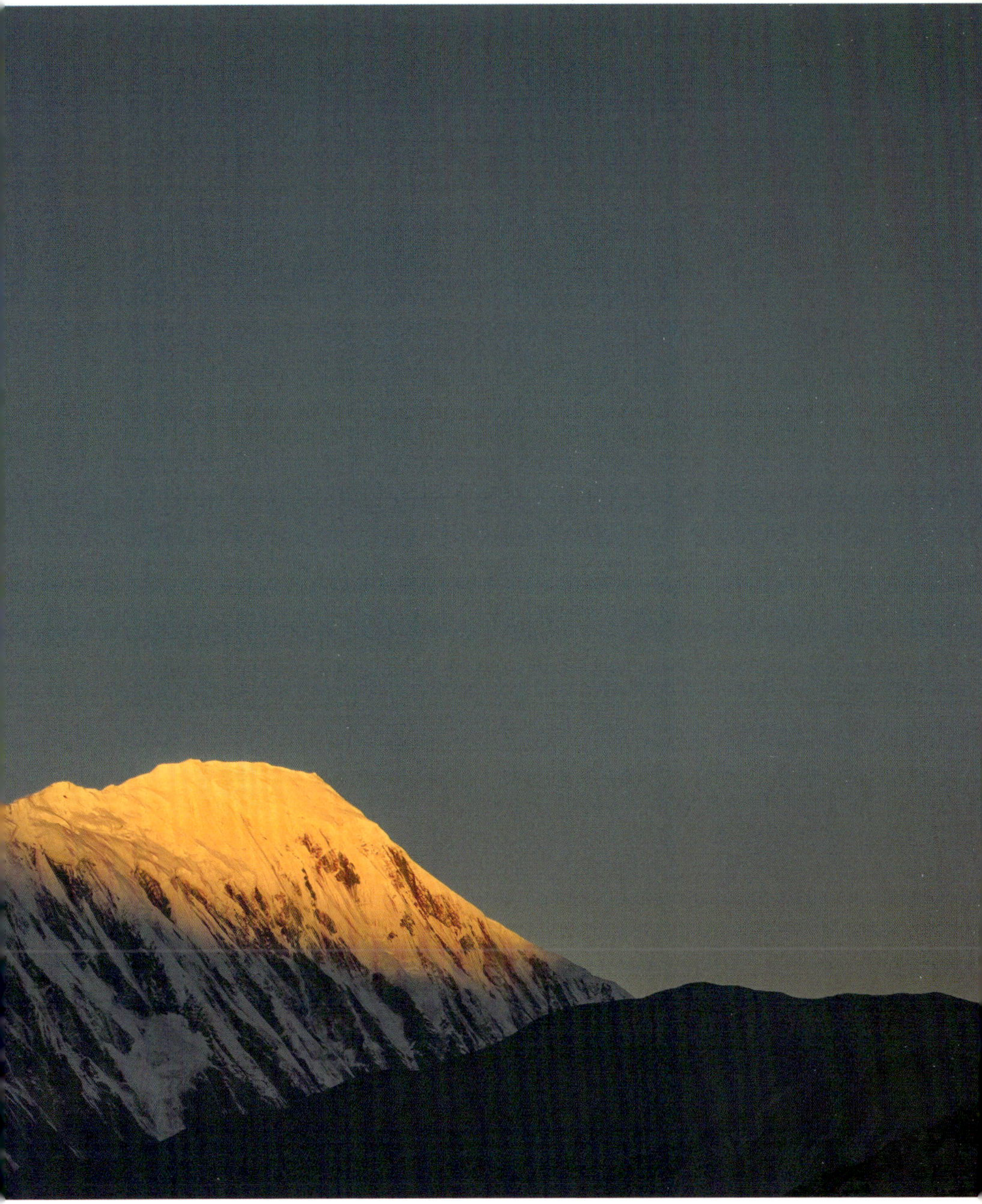

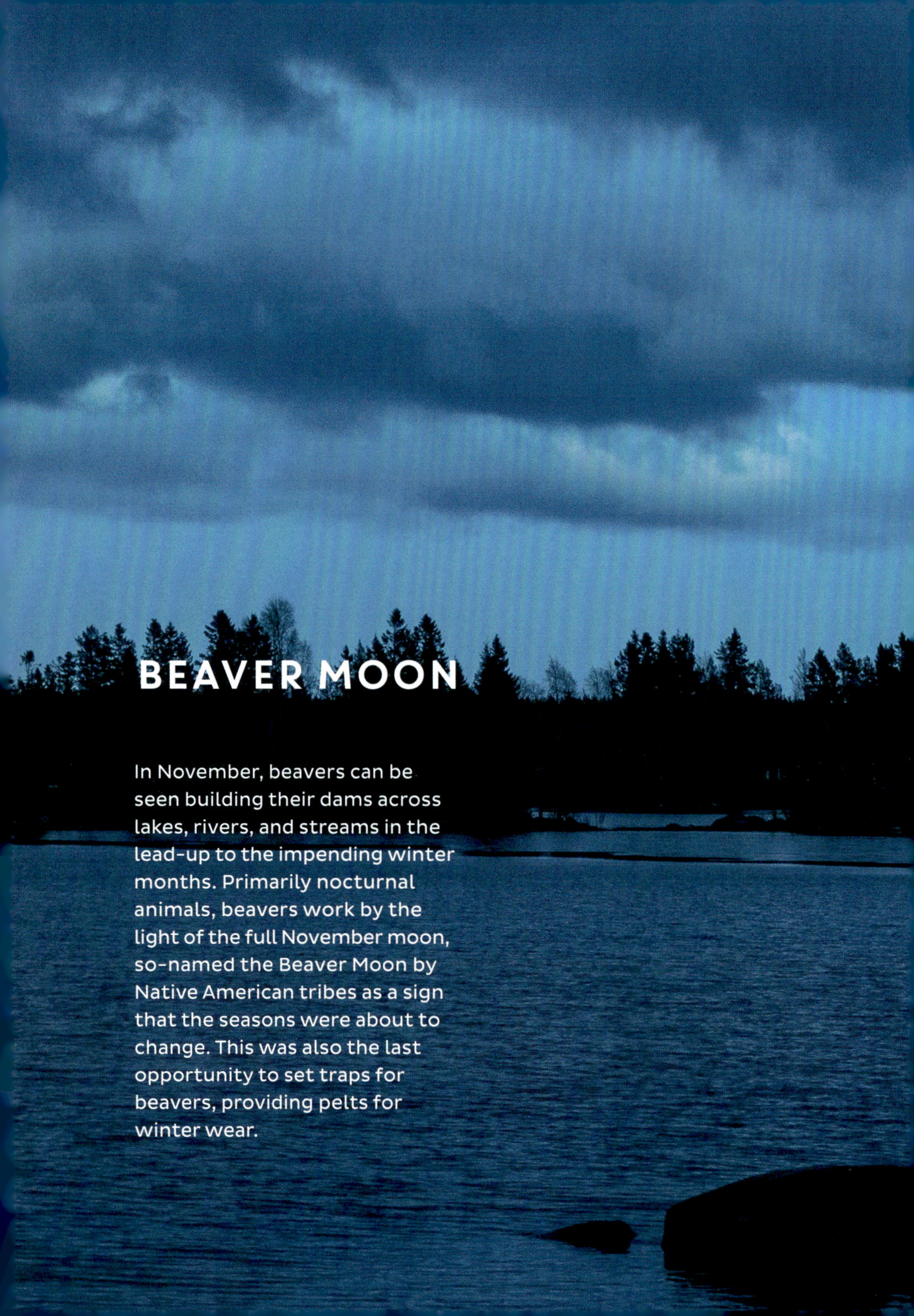

BEAVER MOON

In November, beavers can be seen building their dams across lakes, rivers, and streams in the lead-up to the impending winter months. Primarily nocturnal animals, beavers work by the light of the full November moon, so-named the Beaver Moon by Native American tribes as a sign that the seasons were about to change. This was also the last opportunity to set traps for beavers, providing pelts for winter wear.

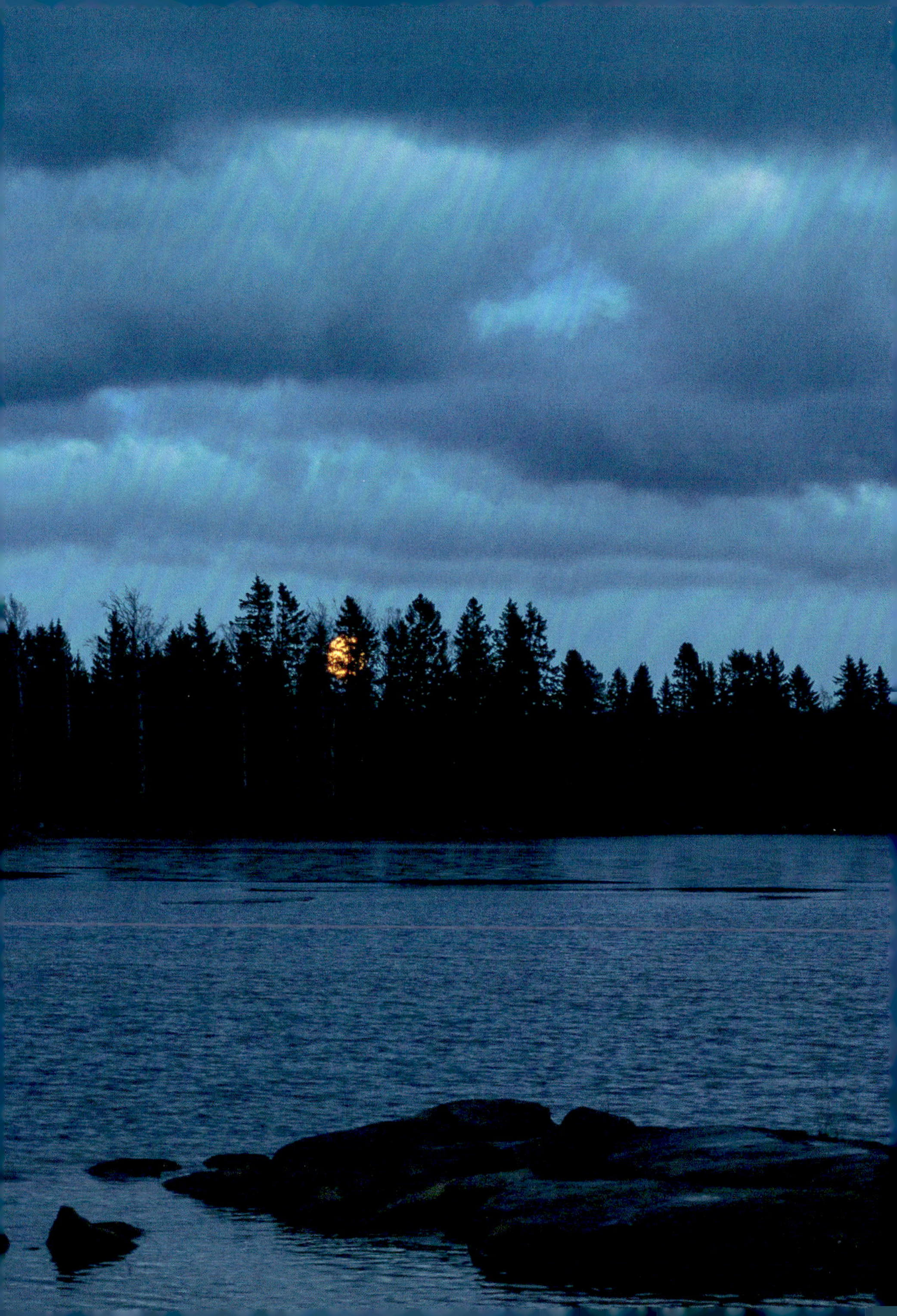

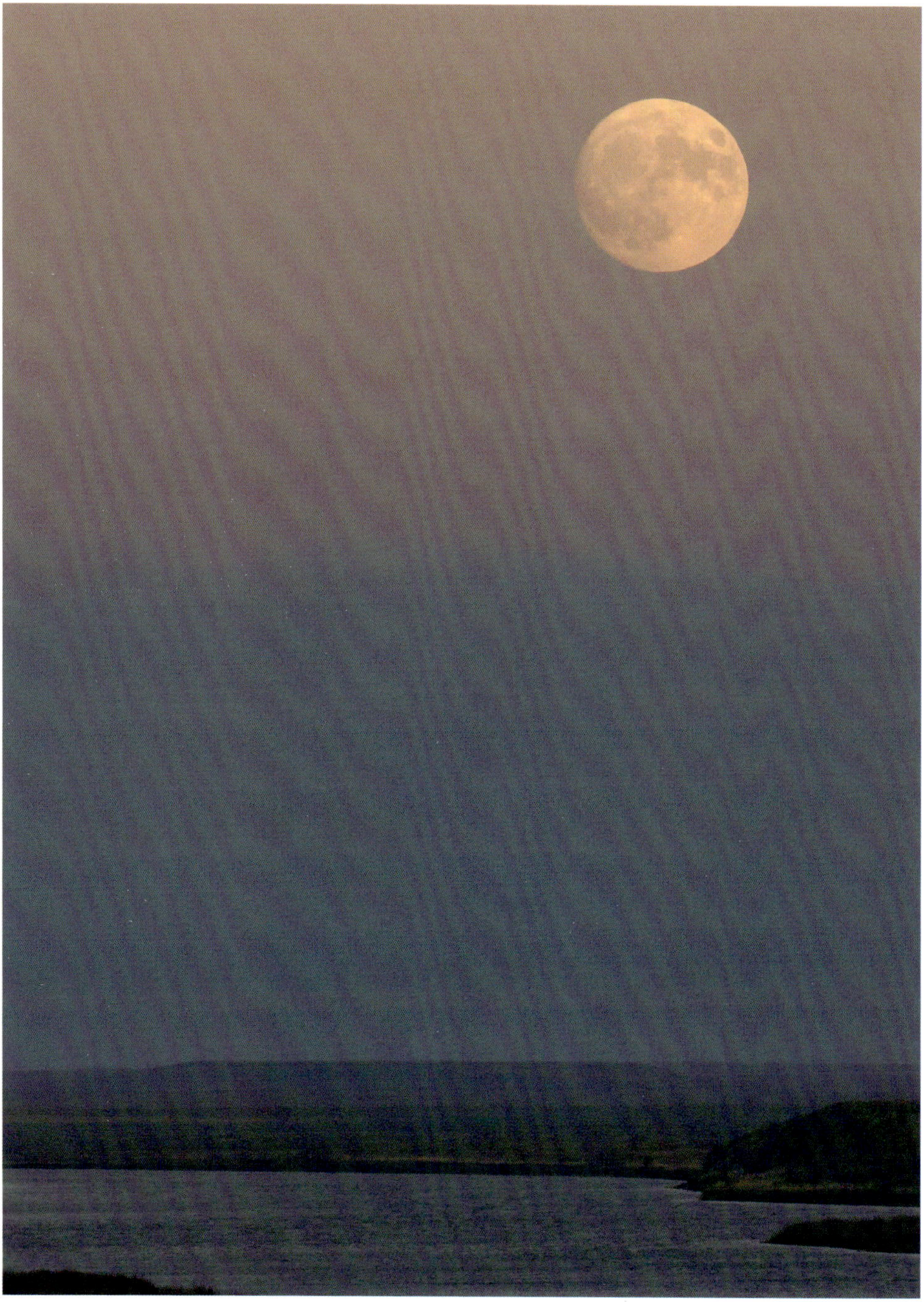

FREEZING MIST MOON

In Inuit culture, the November full moon was called the Freezing Mist Moon. The name is said to be derived from the experience of the Inuit people opening their doors and having freezing white mist fill their igloos, a sure sign of the arrival of winter.

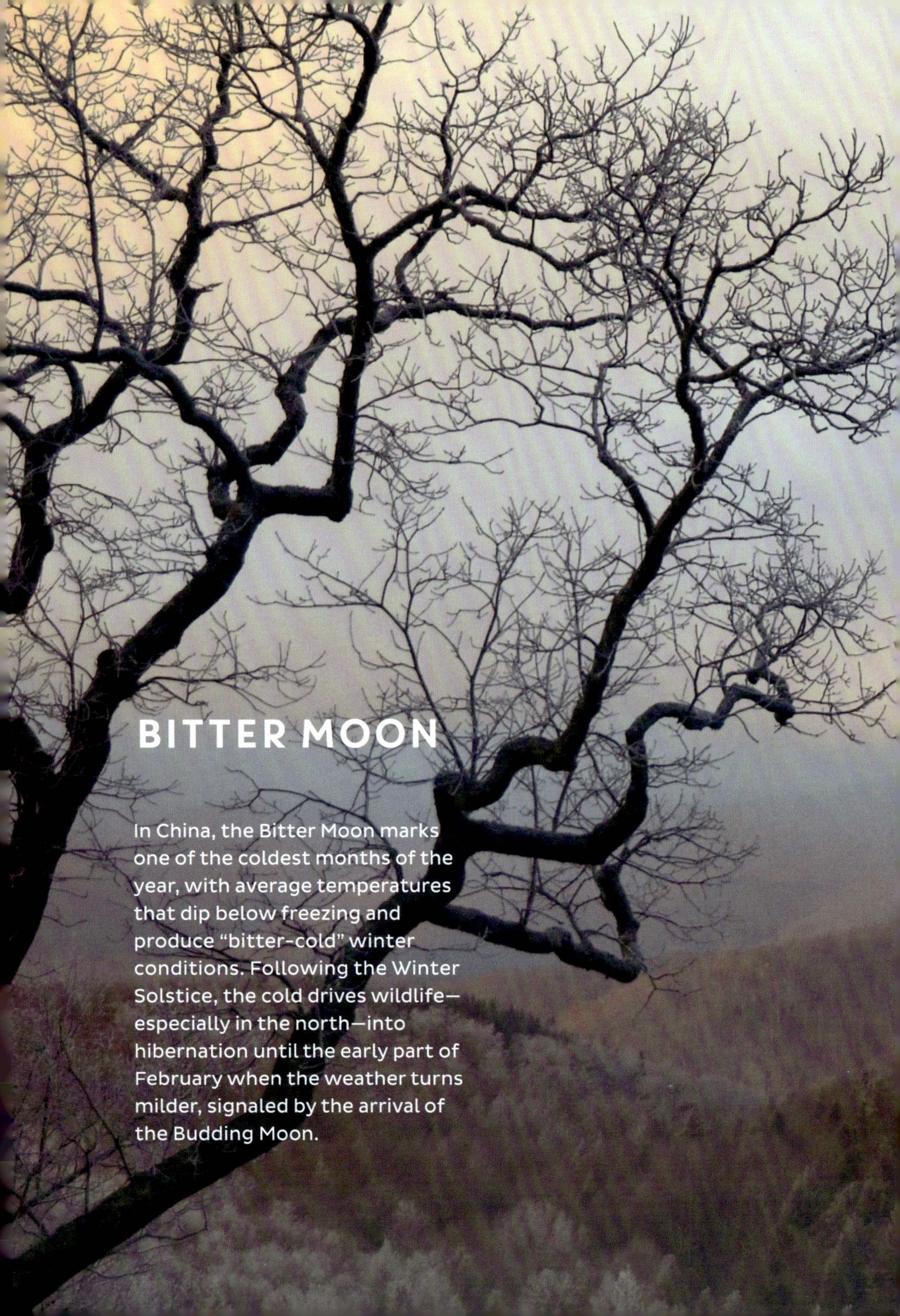

BITTER MOON

In China, the Bitter Moon marks one of the coldest months of the year, with average temperatures that dip below freezing and produce "bitter-cold" winter conditions. Following the Winter Solstice, the cold drives wildlife—especially in the north—into hibernation until the early part of February when the weather turns milder, signaled by the arrival of the Budding Moon.

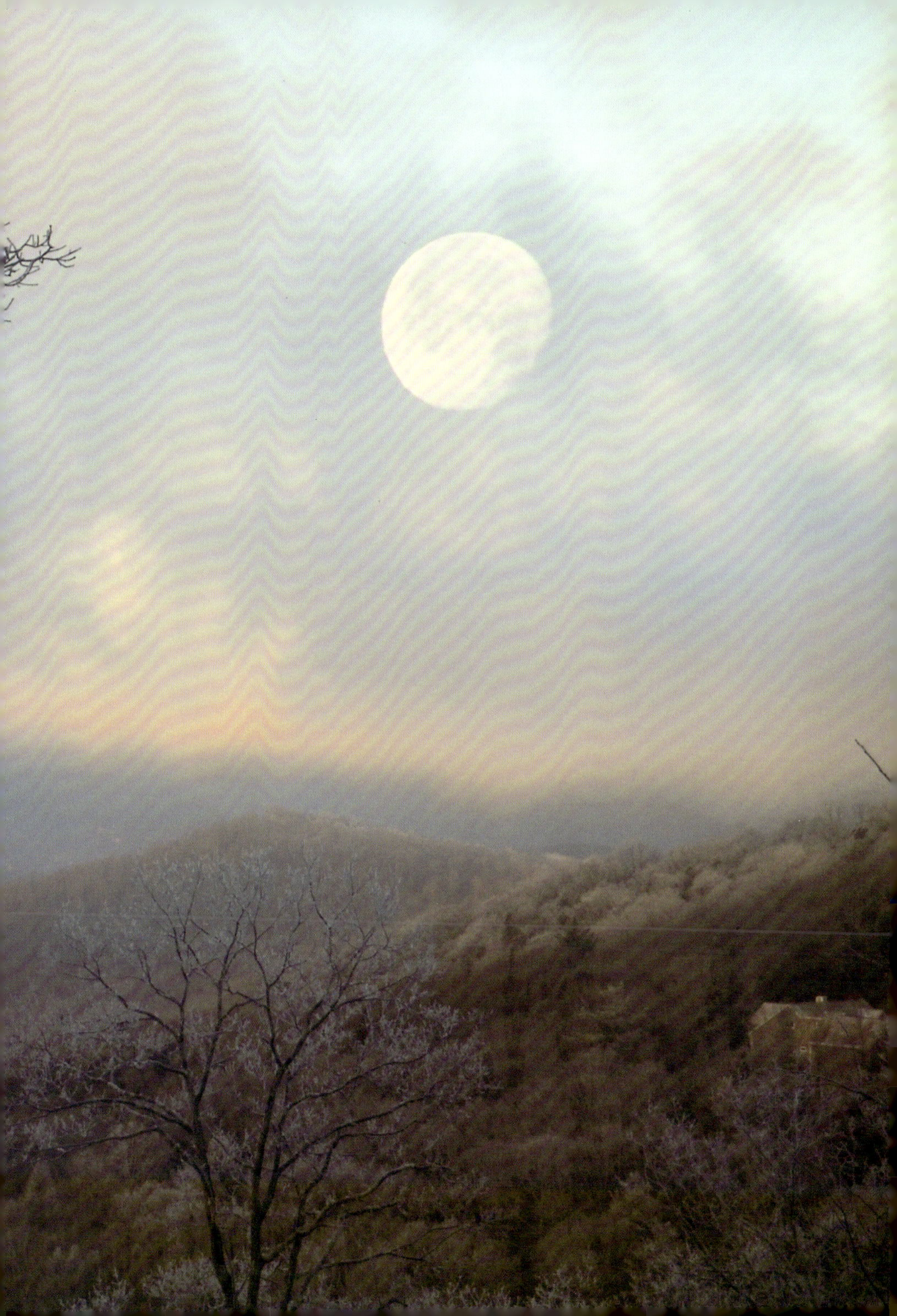

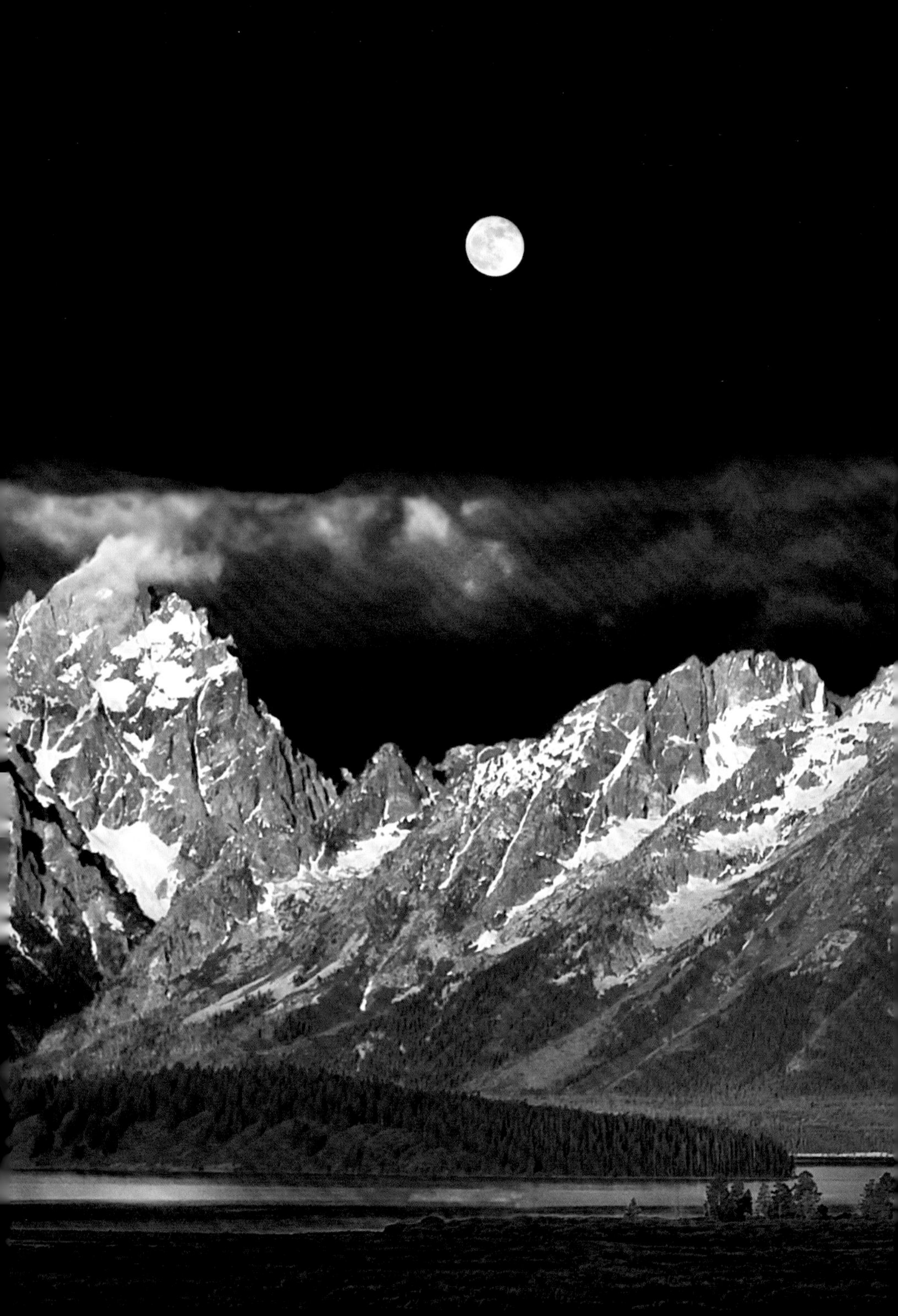

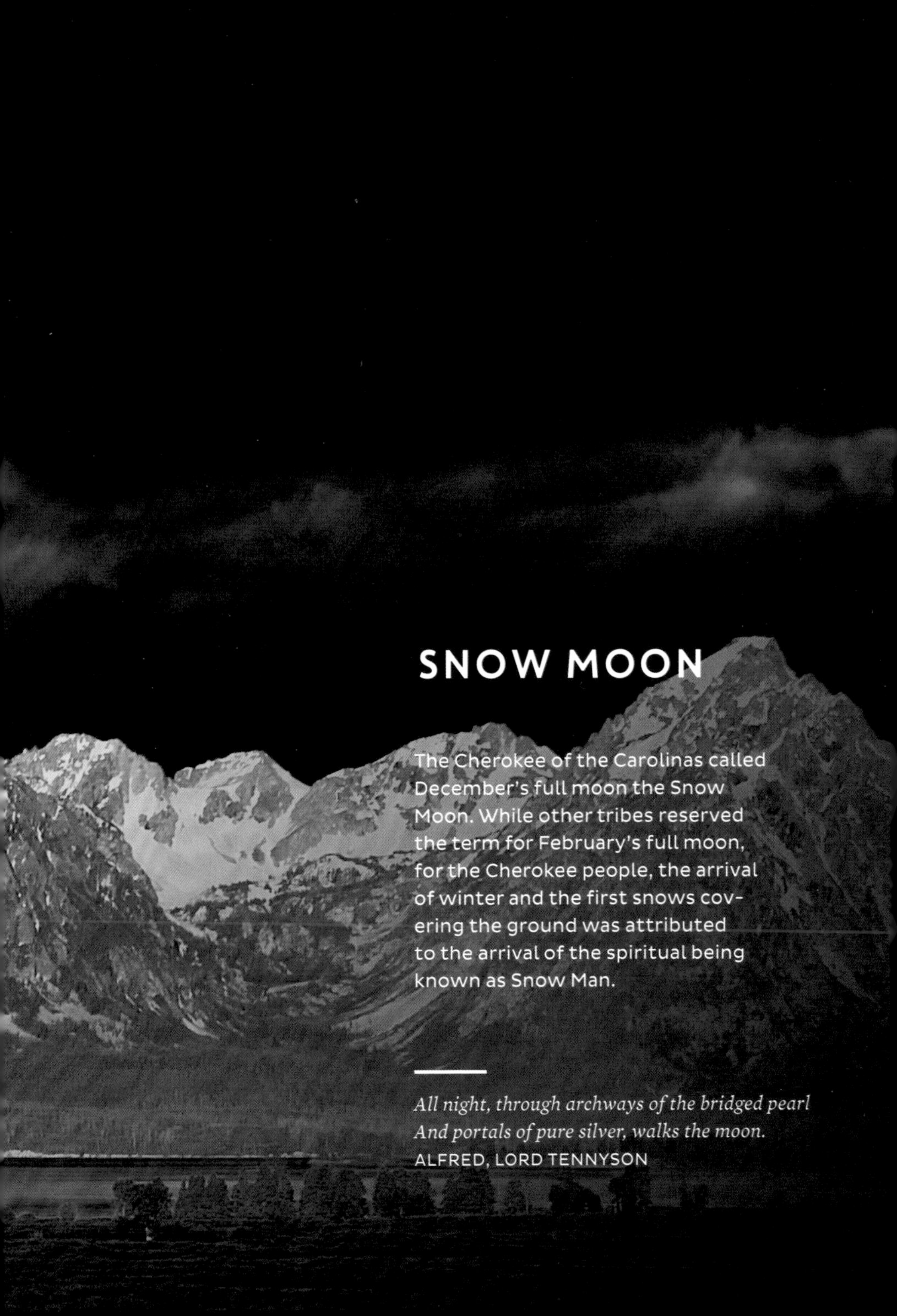

SNOW MOON

The Cherokee of the Carolinas called December's full moon the Snow Moon. While other tribes reserved the term for February's full moon, for the Cherokee people, the arrival of winter and the first snows covering the ground was attributed to the arrival of the spiritual being known as Snow Man.

All night, through archways of the bridged pearl
And portals of pure silver, walks the moon.
ALFRED, LORD TENNYSON

OAK MOON

Some pagan traditions name December's full moon in tribute to the sacred oak. Druids worshipped in oak groves, paying homage to the tree—a symbol of durability and constancy—and a source of strength to cling to in the winter months. It is thought that the name "druid" is an amalgamation of the Celtic word for oak, *duir*, and *wyd*, which means "to know."

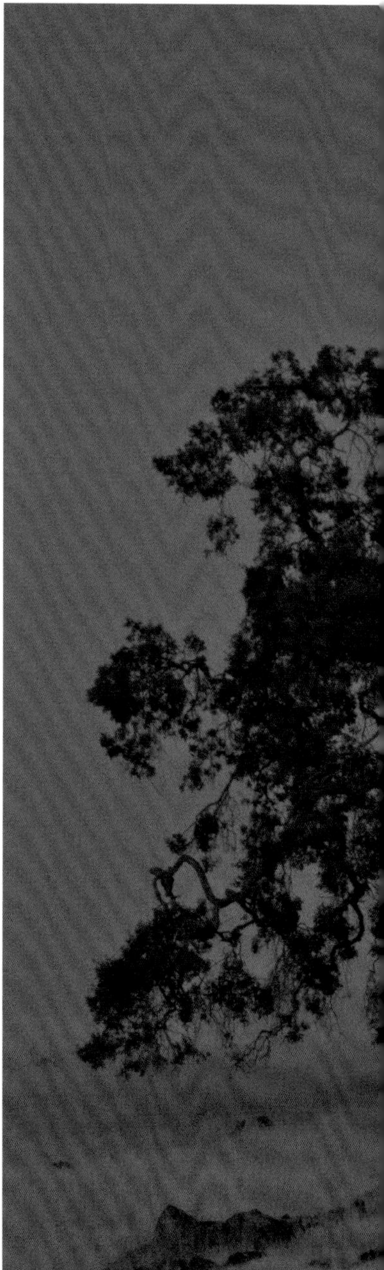

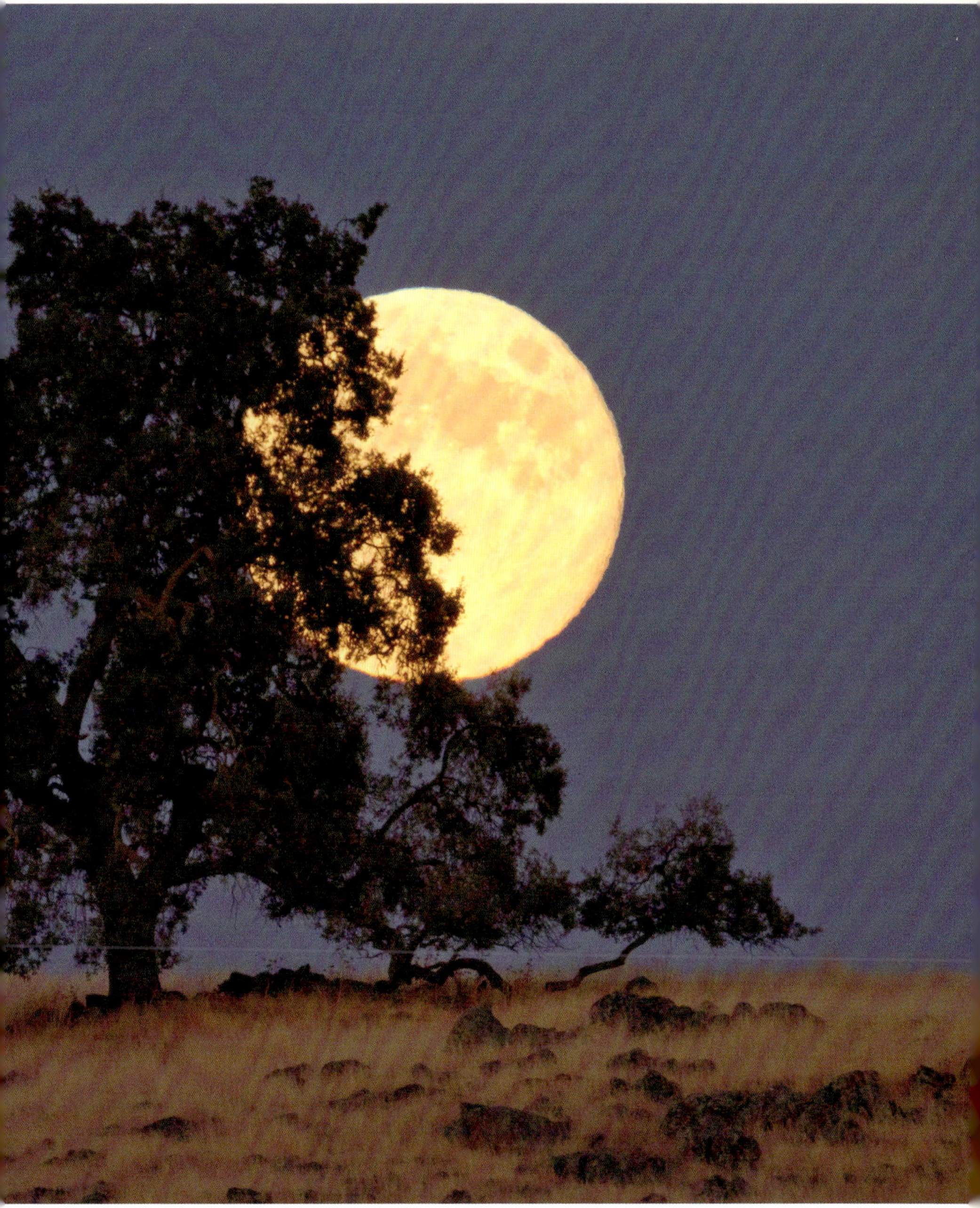

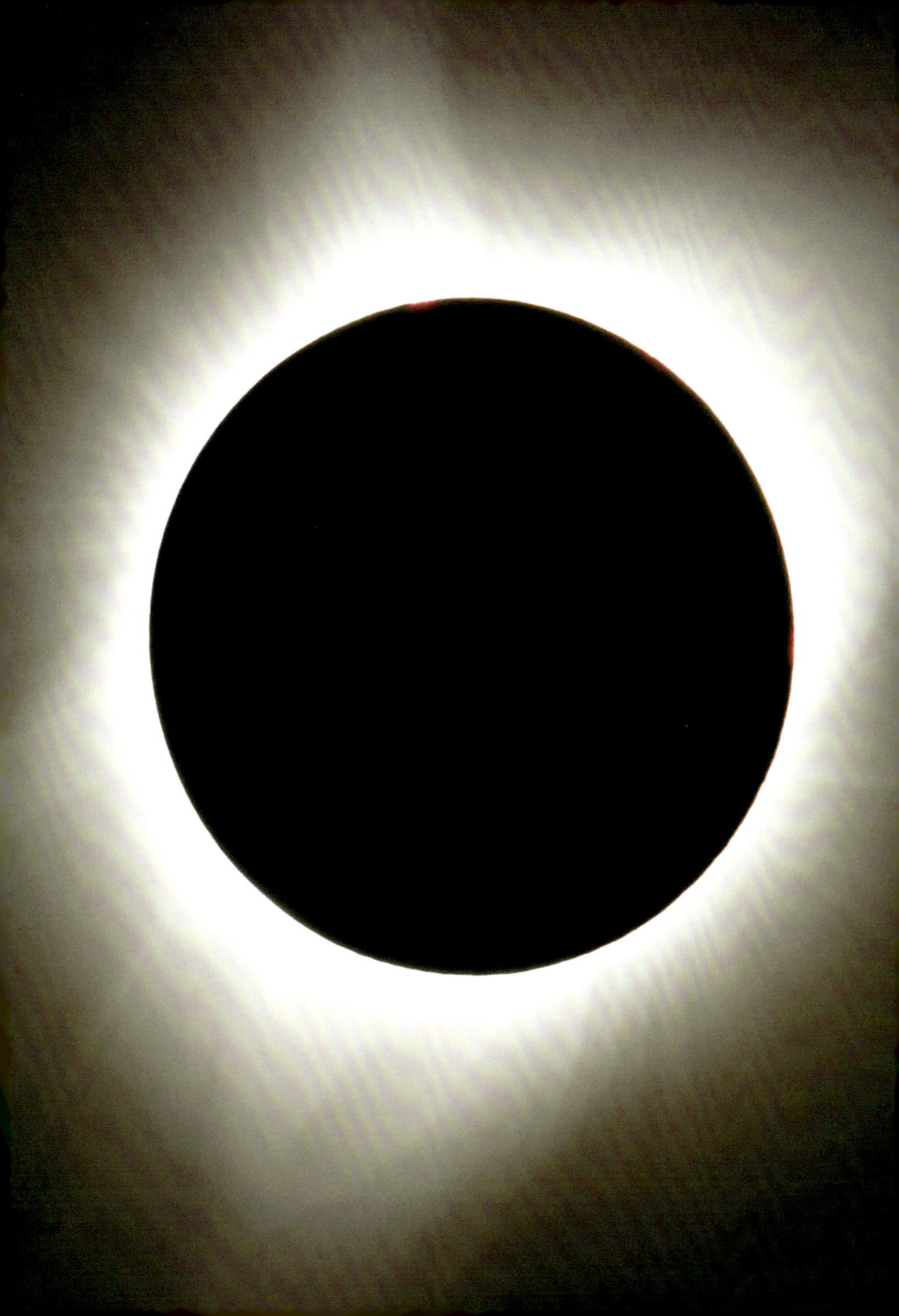

BLACK MOON

There is no one definition of a Black Moon. Usually, it refers to the phenomenon of an additional new moon in a calendar month or season, or the absence of a full moon in a month. The latter can happen only in February, and thus is a rare event—occurring about once every nineteen years.

In many pagan traditions, the Black Moon was thought to be a propitious time to conduct rituals or to cast spells, marking a time of new beginnings and potential. Today, many practitioners of Wicca still hold this belief.

BLUE MOON

The term Blue Moon is often used colloquially to refer to the second full moon in a single month, but older traditions refer to a seasonal Blue Moon: the appearance of the third of four full moons in a season, or an extra thirteenth full moon in a year that would normally have twelve. While the former is somewhat common, depending on time zones, the latter is much rarer—occurring once every two or three years—thus the phrase, "once in a blue moon."

The term Blue Moon has nothing to do specifically with the moon's coloration; although, due to atmospheric conditions such as particles from volcanic eruptions or fires, the moon may sometimes appear to have a bluish tint—a blue moon, but not a Blue Moon.

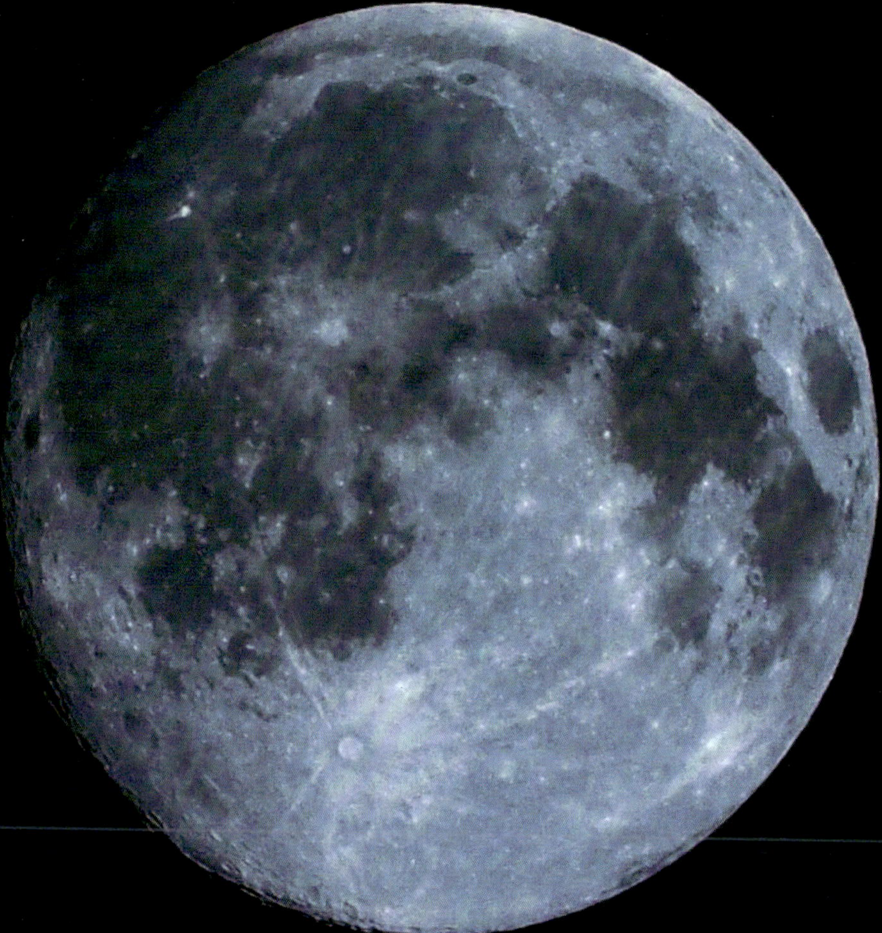

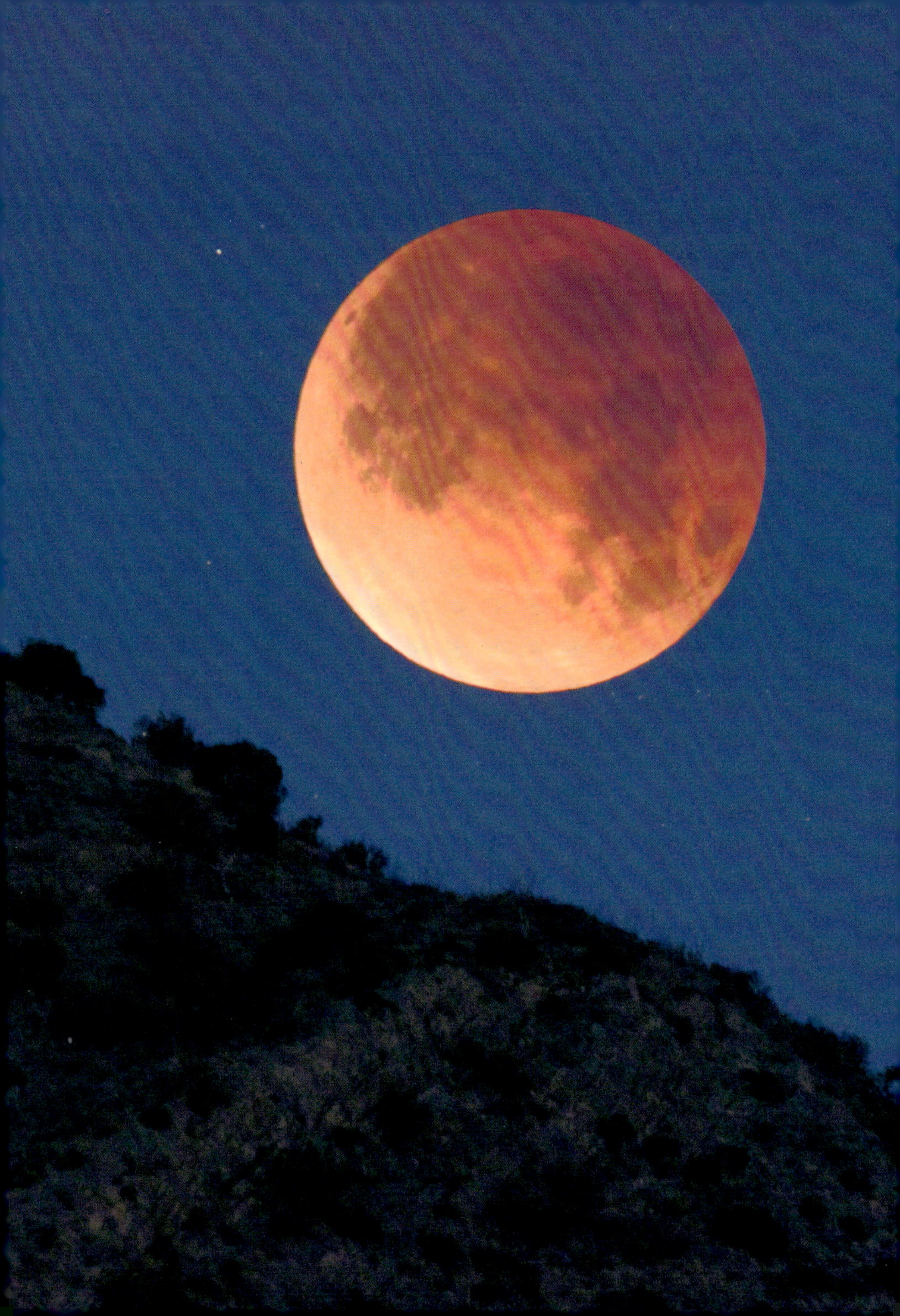

BLOOD MOON

This term is applied to the phenomenon of a total lunar eclipse, when the Sun, Earth, and Moon align so that the shadow of the Earth on the Moon causes it to take on a rich red hue. Because this astronomical event is rare, and its cause has only recently been revealed in the scope of human history, it has been regarded through the ages with awe and foreboding.

PHASES OF THE MOON

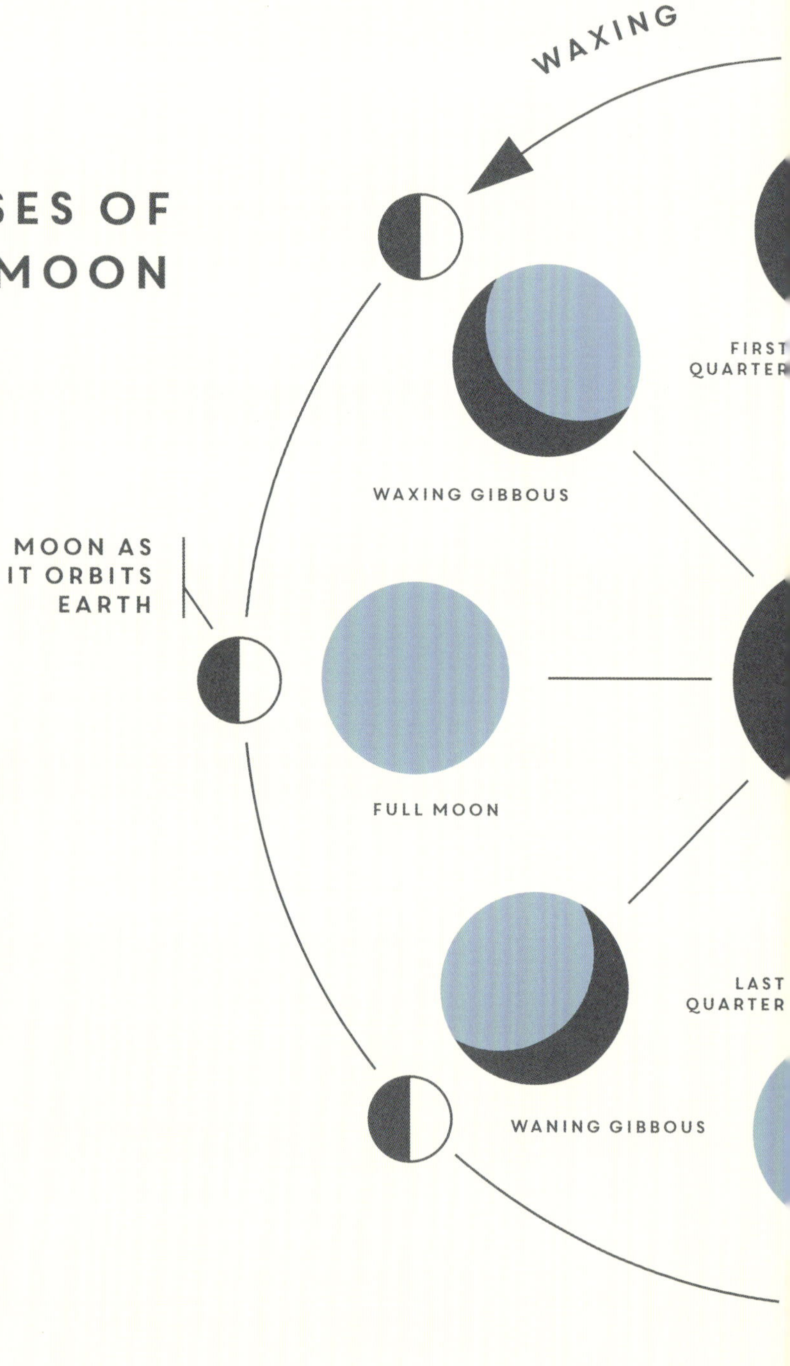

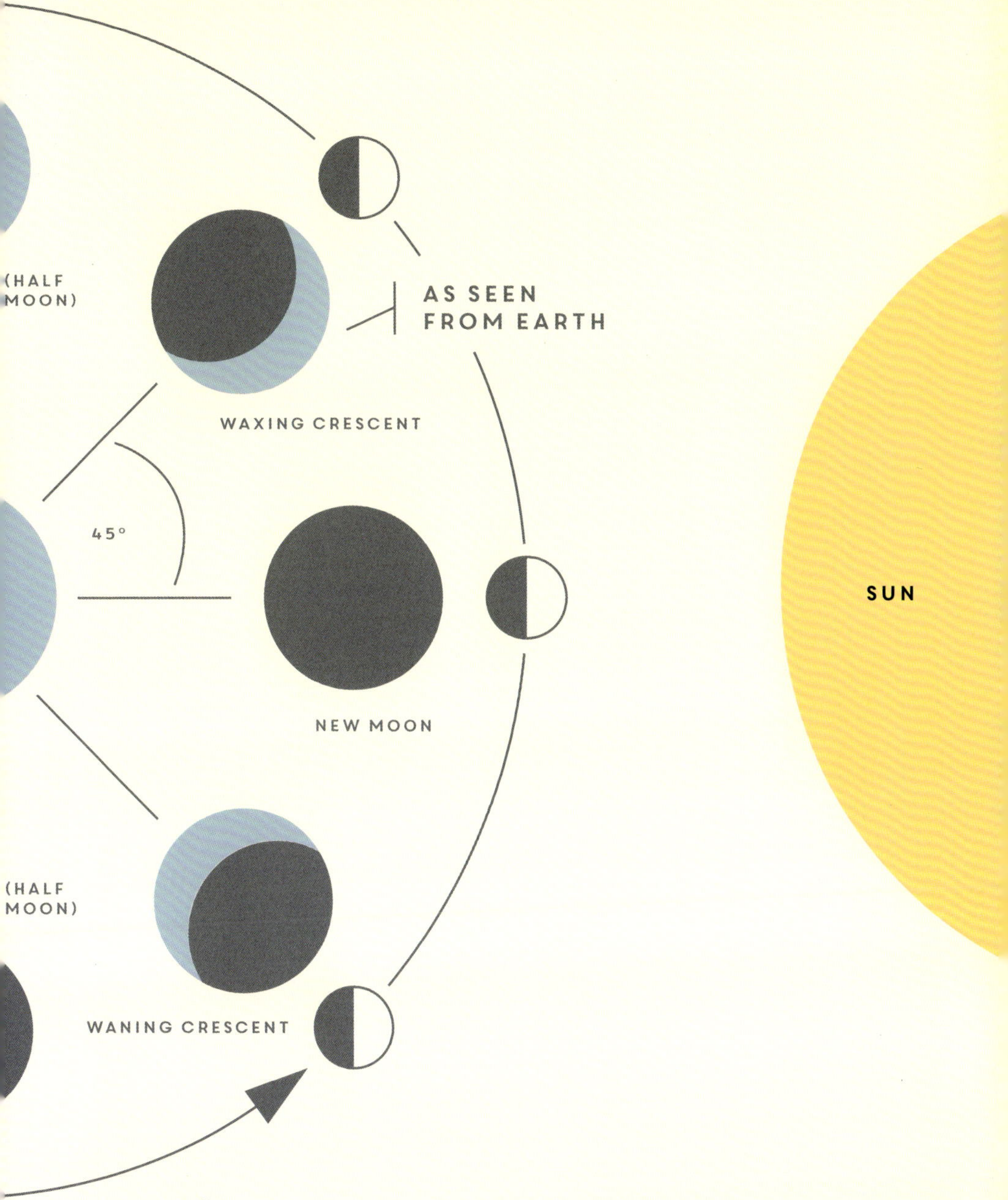
(HALF
MOON)
AS SEEN
FROM EARTH
WAXING CRESCENT
45°
NEW MOON
SUN
(HALF
MOON)
WANING CRESCENT
WANING